THE NEW FARM

THE NEW FARM

OUR TEN YEARS
ON THE FRONT
LINES OF THE
GOOD FOOD
REVOLUTION

BRENT PRESTON

ABRAMS PRESS, NEW YORK

First published by Random House of Canada in 2017.

Library of Congress Control Number: 2017949747

ISBN: 978-1-4197-3108-2
eISBN: 978-1-68335-302-7

Printed and bound in the United States
10 9 8 7 6 5 4 3 2 1

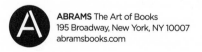

ABRAMS The Art of Books
195 Broadway, New York, NY 10007
abramsbooks.com

This book, like our farm, is the result of a deep and enduring partnership with my beautiful wife, Gillian Flies. The story is told from my perspective, but it is our story.

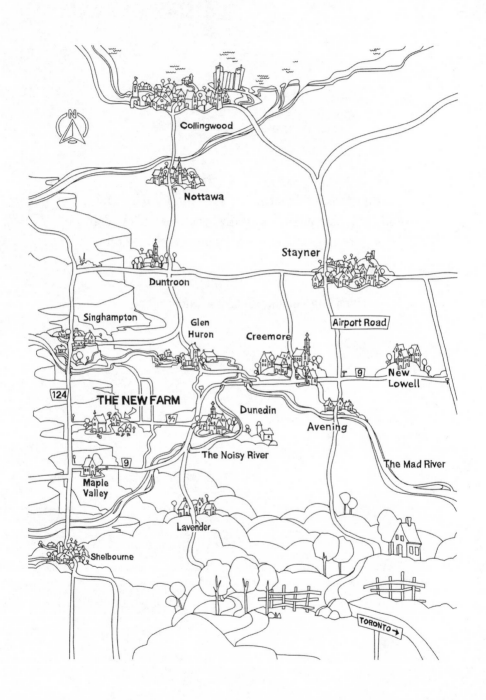

CONTENTS

EUTHANASIA FOR DUMMIES

ALL HAPPY CHICKENS ARE ALIKE; each unhappy chicken is unhappy in its own way.

It all has to do with the coop. If a chicken's coop is too small, the chicken will be pecked and harassed by its coop-mates. If its coop is too damp, it might catch the flu and die. If its coop is too cold, it will get frostbite on its comb and the comb will start bleeding. Worst of all, if a chicken's coop isn't properly sealed, varmints will slip in during the night and tear it to pieces. But when I woke on that fine July morning back in 2005, I wasn't worried about any of that. I had built a beautiful coop. I knew my chickens would be happy.

It was a Saturday, and still early. Gillian and the kids were asleep, so I got up quietly and eased myself down our squeaky stairs. Outside it was cool, calm and silent, and there wasn't a cloud in the sky. In the backyard I stopped and surveyed my domain: flat, green farmland stretching off in every direction, an unbroken ring of forest on the horizon. I had that feeling I sometimes get when I'm up by myself in the early morning, that

I'm lucky to be in this place, at this moment. We had been living on the farm for almost a year, and I was suddenly struck by the wondrous realization that we actually owned this place, that all this land was ours—a realization that struck me on a regular basis back then, and sometimes still does.

In a single year Gillian and I had gone from being relatively normal urban professionals living in downtown Toronto to owners of a hundred-acre farm outside the village of Creemore, about a two-hour drive northwest of the city. We had been taken by the aspirational dream of living in the country, but like many actual dreams, this dream was fuzzy and vague and didn't make a whole lot of sense if you thought about it too much. We wanted to raise our own food, to have animals and a big garden, but we didn't really know what we were doing. We were enthusiastic and idealistic and profoundly naive. If our current selves could meet the people we were back then, we would look on ourselves with a mixture of pity and amusement.

I had woken up early that morning in a state of excited anticipation. Our chickens had just spent their first night in their new coop, a structure that I had spent an inordinate amount of time thinking about, designing and building. I had spent so much time that the coop wasn't finished when our chicks arrived, so they spent their first week of life in our bathtub, in the bathroom next to the kitchen.

I didn't know anything about chickens when we first moved to the farm. But we wanted livestock, and chickens seemed like the obvious place to start. I had to learn a whole chicken nomenclature in the beginning. Chickens bred to lay eggs are "layers" or "laying hens." Chickens bred for meat—"broilers" or simply "chickens"—are very different animals. We had been persuaded by the hatchery's website to order Special Dual Purpose chickens that supposedly combined the best attributes of layers and broilers. The website told us that White Rock broilers, the overwhelmingly dominant meat chicken variety, had been bred into such freakishly efficient gainers of weight that they couldn't walk properly and were susceptible to all sorts of diseases if raised without antibiotic-laden feed (a claim that we later found to be false). If we were aiming for free-range and organic, we were assured, the Special Dual Purpose was the bird for us.

Our batch of day-old chicks, fifty of them, arrived at our local feed store in a very small cardboard box. I began learning new things about chickens at a rapid rate. Day-old chicks are tiny and fluffy yellow and incredibly cute, but they are also very loud, mobile and assertive. They chirped loudly when the kids picked them up or when they were hungry (which was pretty much all the time) or seemingly just for the hell of it, all day long. We put down wood shavings in the bathtub and hung a three-hundred-watt heat lamp over them to keep them warm. It occurred to me that many chickens' lives are bookended by heat lamps. Even Special Dual Purpose chickens have been bred to rapidly put on weight, so the chicks had an insatiable appetite. From day one they would frantically climb over each other to get at their food, and they also drank a lot of water. I would fill up their little feed trough immediately before going to bed, but they would eat it all during the night. Our bedroom was on the second floor, but their manic chirping was loud enough to wake me before dawn.

Any animal with such a rapid metabolism produces a lot of waste. Our chicks were shit-producing machines (birds don't urinate, another thing I learned early on). The bathroom rapidly became very hot, very humid and indescribably smelly. It was like some sort of dystopian sauna in there, and the stench began to pervade the whole house. Gillian let it be known that my coop construction should be expedited.

It takes about ten minutes on gravel roads to drive from our farm to Hamilton Brothers, considered by many (or at least by me) to be the greatest retail establishment on the face of the earth. Hamilton Brothers is a farm and building supply store, but it sells almost everything. I once left there with some plumbing supplies, a box of ammunition, 250 feet of bungee cord and a flat of eggs. I kept the handwritten receipt as a souvenir. It's also the place where I bought my coop-making materials.

To say Hamilton Brothers is old school would be a serious understatement. I have never seen a computer anywhere on the premises, though they must have one in a back office somewhere, because the statements I receive in the mail appear to be created on a dot-matrix printer. Its many separate buildings and yards make up about half of the tiny village of Glen Huron, tucked under a dam at the head of the narrow Mad River valley. The river still powers the Hamilton Brothers feed mill, a five-storey steel-clad building filled with cobwebs, wooden chutes and giant drive belts that towers over the building supply store and the main lumberyard. Across the street is the farm supply building, and behind that is the welding shop and a big hangar where they keep the sheet metal, concrete mix and drywall. Around the corner and past a few houses is another building with tongue-and-groove flooring and fence posts, and across from that is a second lumberyard, for all the pressure-treated stuff. When you call Hamilton Brothers, you have to ask for either the building side or the farm supply side, depending on what you're looking for. The gas and diesel pumps are on the farm supply side. The staff sometimes travel from building to building by bicycle.

When I first started making trips to buy coop supplies, I was accustomed to the anonymity of big-box building supply stores, where I would pile everything I needed onto a giant cart and haul it out to my car without speaking to anyone. But at Hamilton

Brothers, not much is self-serve, and I was forced to interact with the guys at the counter. These were all middle-aged men who evidently knew a lot about everything. They would ask me questions about my order that I often couldn't answer. "What size chicken wire?" or "Ardox nails or regular?" or the one that always struck me with fear, "What you doing with all this stuff?" It seemed they were running my order through a vast mental database and determining that there was nothing known to humanity that could be built properly with the list of items I wanted. I was terrified of looking like an idiot, so I would blurt out an answer and end up back at the farm with the wrong thing, and be forced to return the next day.

After dozens of trips, one of the guys finally took pity on me and took me under his wing. Ivan is a giant of a man, probably six foot five, with massive hands and a huge head. I confessed to him that I was building a chicken coop but had no clue what I was doing. Ivan took me up into the loft above the store where the chicken wire was kept and helped me choose from the surprising range of options—rolls of different lengths and widths, with different size holes and different gauges of wire. After a while he would break into a broad grin whenever he saw me come in. "That must be some chicken coop you're building!" he'd say. I promised to bring him a bird in the fall.

One of the reasons the coop was such an undertaking (beyond simple incompetence) was that my father-in-law's warnings about varmints had put the fear of god into me. Gillian's parents had raised chickens and other birds on their farm in Vermont since the mid-sixties, and Dick warned me that there is almost nothing that won't eat a chicken if given the chance—skunks, foxes, coyotes, all members of the weasel family. Raccoons, he said gravely, will kill your whole flock just for the sport of it, and leave without eating anything. The salient point when it comes to coop construction is that varmints will find and exploit any weakness, no matter how small. Chicken-loving predators share a special evolutionary adaptation, Dick explained: the ability to squeeze through tiny cracks in any enclosure.

But the coop I had constructed was solid, tight and completely varmint-proof. It was built into a lean-to that had been added to the east side of our classic Ontario bank barn, a majestic structure constructed of massive beech beams that was at least a hundred years old. I had framed in a section of the lean-to that was about thirty feet square and then sealed the whole thing with chicken wire, all the way up to the roof. Every corner was meticulously tacked down, and the overlapping runs of mesh were wired together at the seams. I had built a sturdy door that locked with heavy slide bolts, leaving a gap at the bottom of less than a quarter of an inch. Nothing could get in there that didn't have opposable thumbs. So I was confident as I strode across the lawn that fine June morning.

I could see the chicks through the chicken wire as I approached the coop, sleeping in a pile right up against the mesh in a corner. They jumped up as I got closer and started running about in the straw on the concrete floor. Chicks are ridiculously cute for the first five days of life. Then their yellow fluff begins to give way to tufts of white adult feathers and they take on a mangy, adolescent look. At a week old, my birds had bodies about the size of a baseball, with spindly legs and heads half covered with moulting chick fluff. They were ugly.

I hooked my fingers in the wire and peered in at the flock, and a wave of relief washed over me. Everyone appeared to be fine. But as I stood there longer, I noticed a pile of white feathers near the front corner of the coop. I looked down and saw more white feathers outside the wire, near my feet. What was going on? I scanned the flock again. Every chicken was up and moving. None of them was dead. Could a varmint have entered the coop, killed a chicken and then removed the body? I couldn't accept that this was possible. I had sealed that coop.

Then I caught a glimpse of something strange: one of the chicks appeared to have something wrong with its wing. I opened the door and stepped into the coop. This caused the birds to run around frantically, making them harder to observe. I knelt down and let them settle. As the funny-looking chick walked toward me, I realized that there wasn't something wrong with its wing; its wing was completely gone. I felt a little sick. Another chick ran up and turned broadside in front of me. It's left wing was also missing.

It's hard to keep track of fifty birds as they move about in an open coop, but after a few minutes I determined that five chicks had been injured. The most disturbing part of the whole scene was that none of the wounded birds seemed distressed. They were mangled, but they were acting completely normal—*The Walking Dead*, chicken-style.

I retreated to the house, not sure what I should do and completely mystified as to what had happened. (We eventually realized that a feral cat had reached through the wire while the chicks slept.) Gillian was up, and I told her the unhappy news.

Gillian called her parents. Kathi, her mom, answered. She remained completely calm. "You can usually get spray iodine at the farm supply store," she said. "You'd be surprised what a chicken can recover from." I discussed this with Gillian after she hung up. "I don't think your mom realizes what has happened to these birds," I said. I'm no veterinarian, but it didn't seem to me that iodine was going to fix this problem. Gillian explained that her mom had once brewed a bucket of comfrey tea every morning for three weeks and fed it to a lamb that had broken its leg. "The lengths to which my mother will go to save an injured farm animal are not normal," she said.

Gillian persuaded me to call Hamilton Brothers to see if they had anything that could help. I badly wanted to talk to Ivan, but he was a lumber guy and I knew he would tell me to talk to someone on the other side of the road. Mike, the main farm supply guy, answered the phone. We're now good friends, but back then I was intimidated by his deep well of practical knowledge and his ability to point out my ignorance in a way that was both cheerful and unfailingly polite.

I knew if I simply asked for spray iodine he would ask "What for?" So I briefly explained my problem and threw myself on his mercy. There was a long silence, then Mike said, "I usually find a bullet works well for that."

After a period of reflection, I came to the conclusion that Mike was right—the most humane course of action would be to kill the unfortunate chicks. There was a chance that some of them might recover, but it seemed much more likely that their wounds would become infected and they would die slow, painful deaths, no matter what I did. I had long ago made peace with the fact that eating meat requires the killing of animals, but that peace is easier to make standing in a supermarket aisle than outside your own chicken coop.

I strode grimly back out to the barn, knowing what I had to do. My father is a hunter and I had spent time as a kid on my uncles' farms, so I had a little bit of murky experience to fall back on. On the walk out I decided that I would drown my victims, though I'm not sure where that idea came from. I found an old bucket, filled it to the brim with water and grabbed a length of board that would completely cover the top. I walked into the coop, snatched up the first wounded chick I saw and carried it out to the bucket. I tried to handle it as gently as possible. I also tried not to look at it because I didn't want to think too much about what I was about to do. I quickly dropped the chick into the water and slammed the board over the bucket. Then I waited.

There was no scratching, no vibration, no noise caused by the chick desperately trying to escape. I began to breathe more normally and my heart stopped beating in my throat. After several minutes I carefully lifted one side of the board and peered in, expecting to see a limp mass of wet feathers. Instead I saw a chick that was very much alive, floating about two inches below the surface, eyes wide open, apparently holding its breath. Without thinking I grabbed it out of the water and set it on the ground. It shook its feathers and started hopping around, seemingly none the worse for wear. I guess I could simply have left it in the bucket longer and waited for it to drown, but I was so surprised to learn that chickens could swim, and could also hold their breath for longer than I could, that I gave up on the bucket.

When a suburban-raised, middle class, North American male finds himself all alone in a visceral, primordial, life-and-death situation, it's difficult to predict how he will react. My upbringing to that point had provided me with neither the knowledge nor the emotional fortitude necessary to quickly and effectively euthanize those unfortunate chicks. I'll spare you the graphic details, but I resorted to an escalating series of measures that included, but were not limited to, strangulation and bludgeoning with a length of two-by-four. I worked myself into a desperate, anguished frenzy, and I did something that a year earlier I never could have imagined I would do.

At the end of the ordeal I was hyperventilating and sweating profusely, utterly horrified by what I had done. The injured chicks were all dead, and so was the bucolic dream of life on the farm. The dream was dead, almost before it began.

DEAR READER

YOU WOULD BE FORGIVEN, HAVING just read the story of our first experience with life and death in the barnyard, for thinking that this book is yet another in the well-established City Folks Move to the Country, Chaos Ensues genre. You would be forgiven for thinking that Gillian and I, having decided to experience life on the farm, had found it nasty, brutish and ridiculous, and then written a book. You would be forgiven for assuming that we, like so many before us, had moved to the country not to farm but to write about farming. You would be forgiven, but you would be wrong.

Or at least you would be mostly wrong. It's true that when we left our home in downtown Toronto and moved to the farm more than ten years ago, we had no idea what we were doing. We made an enormous number of stupid mistakes in the early years, some of which are pretty funny in hindsight. But our goal was never to buy a few chickens and plant a few seeds, then blog about it for a year or two before retreating to our lives in the city. Our goal, once we figured it out, was to create a *real farm*.

And what is a real farm? It is, first and foremost, a business. A farm can and should be a lot of things: a place where food is grown, a sanctuary for wild things, a gathering place for family and community. But for a farm to endure, for a farm to be sustainable in the broadest sense of the word, it must make money.

For years Gillian and I focused on profitability with an intensity that bordered on obsession. We fought and struggled and eventually achieved our goal, but at enormous cost. Our bodies broke down, we were consumed by stress and we came very close to losing our marriage. It took us several years *after* we had achieved financial success to realize that making money wasn't enough. Profitability was necessary, but not sufficient.

As I write, we have just finished our tenth season of farming. Our business is established, stable and profitable. We have paid off all our debt and the farm is mortgage-free. We have a team of skilled and dedicated employees who share in our profits. We make enough money to save for our children's education and to travel the world in the off-season. Our farm is our sole source of income; our twenty acres of garden provide everything our family needs. Our farm is small and organic, and it's *real*. But it's so much more than that.

Our farm is proof that small-scale, sustainable farming is a viable alternative. It's a place of community, where chefs, activists and foodies gather to plot the overthrow of all things evil: industrial agriculture, the rat race, high-fructose corn syrup. Our farm is the centre of gravity around which we have built a happy, meaningful and productive life for ourselves. And most importantly, our farm is testament to the power of the good food movement to radically change our food system from the ground up.

Right now there may be a voice somewhere deep inside you that's saying, "Maybe one day *I'll* ditch my desk job and move to the country and buy some land. Maybe one day *I'll* start a farm."

The fact that you're reading this book makes me think that some tiny part of you has at least wondered what it would be like to be a farmer. Not so fast. This book says that a new kind of farming is both possible and necessary, but this new kind of farming is also really fucking hard. Before you get too excited, you'd better hear the whole story.

CHAPTER 1

BUYING THE FARM

PEOPLE OFTEN ASK GILLIAN AND ME how we ended up on the farm. It's hard to know how far back to go when answering that question. I grew up in Scarborough, the easternmost and most reviled suburb of Toronto. Mike Myers grew up there, too, and says that Scarborough was his inspiration for *Wayne's World*, which sums the place up pretty nicely. I had a great childhood, but even when I was very young I knew I wanted to get out of there and see the world. Gillian grew up on a sheep farm outside the tiny village of Williamstown, Vermont, and similarly vowed to escape as soon as she could. We launched ourselves into the world at about the same time in the early 1990s, both of us propelled by a potent mix of idealism, ambition, adrenaline and libido, both of us utterly assured of our own immortality.

The inevitable crossing of our paths occurred in 1994. We were both working for the National Democratic Institute, an American organization that sought to spread and strengthen democratic government. I was living in Lilongwe, the sleepy capital of the Republic of Malawi in central Africa. Gillian had just finished a two-year stint in Botswana with the Peace Corps and was the Malawi desk officer at NDI headquarters in Washington, D.C. Email was a new and frightening invention at that time, so our romance started by fax. I would photocopy giant cockroaches and wall geckos and make custom letterhead for my messages to her. She fell for it, of course. I convinced her to relocate to the Malawi office, and the rest is history. We became roommates, then sleeping buddies, then fell in love, in that order.

We lived together for two years in Malawi, then spent time in over a dozen countries, sometimes separately but usually together. We had the good fortune to be in the thick of the action as the world sorted itself out after the end of the Cold War. We helped organize the assembly that wrote a new constitution for Malawi, conducted public opinion research in the tribal areas of Yemen, and organized international observer delegations to landmark elections in Ghana, Nigeria and Indonesia, all while we were still in our mid-twenties. We worked incredibly hard, but it was also incredibly fun.

We would probably still be at it today, but our luck finally ran out in 1999 in East Timor. We were working for former president Jimmy Carter at the time, leading a small team of international observers in the run-up to the UN-sponsored referendum on independence from Indonesia. The Indonesian military had created surrogate militias to intimidate the population prior to the vote, and when two-thirds of the Timorese chose independence, the militias literally started to burn the whole place down.

We got caught by a group of drug-crazed, sword-wielding thugs in the port in Dili, the capital, while trying to evacuate our local staff. Gillian and I escaped in separate vehicles, our translator's father clinging to the hood of my car while his extended family screamed inside Gillian's as she raced along behind me. After a high-speed chase through the streets of Dili, a motorbike darted in front of me, the passenger on the back trying to aim a pistol at my head as I swerved. I finally ran down the motorbike to avoid being shot.

We ended up being arrested by the Indonesian police, forced to pay compensation to the guys on the motorbike who had been trying to kill us, escorted to the airport and placed on the last civilian evacuation flight out of East Timor. There's nothing like staring down the barrel of a Colt .45 to shift your focus to the important things in life, one of which is actually staying alive. After East Timor, we moved home and got married.

Gillian and I bought a house in downtown Toronto with my brother Mike, a big old place on Palmerston Boulevard, the most elegant street in the city. Mike and his future wife, Sue, had the top-floor apartment, Gillian and I were on the second floor, and we rented out the first floor and basement apartments. We could walk five minutes south to the bars and restaurants on College Street, the heart of Little Italy, or three minutes north to the sushi and barbecue places in Korea Town, along Bloor Street. Gillian found a job at a public-sector management consulting firm and I started working as a producer for CBC Television, the national broadcaster. Living in a big North American city after ten years in the developing world gave us a heightened appreciation for how good we had it. For a time, just the fact that the roads were smooth, water came out of the taps 24/7 and no one ever tried to kill us was enough.

It didn't take long, however, before we became restless. The first issue was children. The prevailing narrative around the arrival of kids is one of revelatory joy and happiness, the sudden realization that everything has changed. This held true for Gillian and me, but it also became clear that having kids was a major impediment to doing many of the things we most enjoyed doing in the city. What's more, the continued enjoyment of those things by neighbours was affecting our new lives as sleep-deprived parents. It's great to live in a place where you can stumble home from the bar at three in the morning. It's not so great to live in a place where you are frequently woken up by other people stumbling home from the bar at three in the morning, just after your screaming infant has fallen asleep. Foster arrived in 2001 and Ella in 2003. They were both an indescribable joy from the very beginning, but having two kids under two in a two-bedroom apartment can really take the shine off urban living.

The second issue was the rat race. Perhaps a decade of immersion in other cultures had given us the ability to see the flaws in our own more clearly, or perhaps we just weren't the kind of people who could put up with working for the man. Either way, we had trouble accepting the idea that donning a suit and sitting in an office for ten or twelve hours every day was a good way to spend your life. It didn't help that both of us had horrible bosses. It wasn't long before I started spending a large portion of the workday sitting at my desk, quietly plotting my escape.

WE FLED THE MADNESS OF our child-infested apartment and the tedium of our office jobs by leaving the city on weekends. My parents had bought an empty piece of land about two hours northwest of Toronto, and that summer the extended family was building a timber-frame cabin on it. Gillian and I would pack up the kids and fight the Friday afternoon traffic, then spend the weekend camping out, building and drinking beer. We stayed in an old tent trailer, there was nowhere to bathe and the only facilities were an outhouse with no door, so it wasn't particularly relaxing. But once the cabin was done, we thought, we would have a weekend refuge.

In the fall of 2003, the cabin was finally completed, and we were getting ready to spend our first weekend inside. It had no electricity or running water, but at least it had a roof, a couple of sleeping lofts and some decent beds. Ella had been born in June and Gillian was still on maternity leave, so she had the car all packed up and ready to go. I was just finishing up at work when I got a call from my father.

"You can't sleep in the cabin," he said abruptly.

"Why?"

"Some prick from the township just called and said he would fine me a thousand dollars a night if anyone sleeps in it."

My father, it turned out, had neglected to obtain the proper permits before building the cabin. A legal dwelling must have things such as potable water, a septic system and stairways that conform to the building code, requirements that would surprise no one except my father. He had saved himself a bundle of money and paperwork by taking out a permit for a storage shed, a pretty accurate description of what the cabin actually was. But a neighbour had complained, and now our plans for the weekend—and all subsequent weekends—were shot.

When I called Gillian to give her the bad news, she already had both kids in their car seats and was ready to pick me up from the office. "There's no way I'm unpacking everything and taking it back up those stairs," she said. I wasn't about to argue. As we crawled north on Avenue Road, Gillian suggested we find a hotel near the cabin and call a real estate agent. If we really wanted a weekend place, it might be time to look for one of our own.

THE FAMILY CABIN IS LOCATED near the town of Collingwood, at the south end of Georgian Bay, a landscape dominated by the

Niagara Escarpment. The escarpment is a sort of one-sided ridge running from Niagara Falls in the south to the tip of the Bruce Peninsula in the north. In some places the escarpment is a jumble of hills that is difficult to recognize as a distinct landform. In others it's steep and abrupt, with sheer limestone cliffs. Near Collingwood it winds and meanders back on itself, cut with deep river valleys and caves that hold snow into the middle of July. The rolling landscape is a patchwork of beautiful farms and huge tracts of maple forest.

We checked into a hotel in Collingwood and called Vicki Bell, the real estate agent who had helped my parents buy their land. Vicki was born and raised in the area and knows every back road for miles around. We spent all day Saturday driving around with her, the kids squirming in their car seats, Gillian and I becoming increasingly depressed. We thought we could buy an empty chunk of land for a hundred thousand bucks, then slowly build our dream cabin. We were sorely mistaken. Land on the flat, featureless stretches above and below the escarpment was relatively cheap, but also flat and featureless. Places in the hills and valleys of the escarpment itself were crazy expensive. As we drove down the dirt roads, we passed enormous mansions on huge country estates recently built by wealthy weekenders from the city. There was nothing we could remotely afford.

I spent much of the next week poring over real estate listings at work. There must be a deal out there somewhere, I thought. I called Vicki about one place that sounded great—right on the Mad River, with a cool little cottage, for less than $350,000. "I won't even show you that," she said. "It's right behind the Hamilton Brothers' chicken barn." I didn't know why that was such a bad thing, but now I'm glad I listened to her.

Spending too much time on real estate websites can play dangerous tricks on your mind. We had started out looking for a few forested acres to build a modest cabin. After less than a week, we had resigned ourselves to buying a much larger piece of land for way more money, simply because that was all that was available in the area. We probably should have looked elsewhere, or given up on the idea of a weekend place altogether, but we didn't. Instead we got excited when Gillian found a listing for a piece of property—a hundred acres of flat farmland on top of the escarpment with a crappy old farmhouse—that was nothing like what we had set out to find and that we couldn't afford. The weird thing was that Vicki was the listing agent. When we called her about it, she sounded surprised. "Oh," she said. "I didn't think you'd be interested in Rubin's place."

THE NEXT WEEKEND WE LOADED UP the kids once more and drove north. Vicki was busy with other clients but had arranged for the owner to let us in. It was a cold, grey October morning. The leaves were already off the row of old maples that lined the lane leading up to the small, square, two-storey farmhouse. We drove around to the back and parked. There was a big old barn just to the south of the house, and a large steel grain bin in the backyard. A cedar hedge blocked the view to the west, but to the south and east were flat, open fields for just about as far as we could see. The word that sprang to mind was "bleak."

There were no other cars in the drive and no one answered when we knocked at the back door, but it was open, so we let ourselves in. The interior was much like the exterior: cold and desolate. It seemed like the heat had not yet been turned on for the season. Before we could get too far into the house, a pickup arrived outside. A burly man with a scruffy, unshaven face stepped out and came in the door. He paused as he entered to take off his winter boots, revealing a plastic shopping bag on each foot. "I had a bit of a leak," he explained with a grin. I'll never forget my first sight of Rubin McCormick, standing in the empty living room, hand outstretched, feet clad in white plastic bags. "Nice to meet youse," he said. "I'm Rubin."

When I think back to that first tour of the house, I can't figure out why we liked it so much. Rubin had bought the place only a year earlier, moved his family in, then after a few months moved back to his house in the valley, a few miles down the road. His wife didn't like the winters "up on top," he explained. There was no furniture in the house, and it wasn't particularly clean. Rubin's idea of staging had been to pile a huge mound of fresh manure in the barnyard, in a spot that dominated the view from the large living-room window. There was green shag carpet throughout the upstairs; the kitchen had a beige linoleum floor and lime green appliances. One of the bedrooms had some utterly terrifying murals of cartoon characters. I half expected the twins from *The Shining* to step out of a closet. The house was probably a hundred years old, but the interior was straight out of the late 1970s. It smelled mouldy.

As we were readying to leave, I wandered out by myself to take a look at the barn. When I stepped into that huge, empty structure for the first time, I felt like I was entering a sacred place, some sort of ramshackle wooden cathedral. I could see the original cedar shingles under the roof metal forty feet above me, and the adze marks on the massive hand-squared beech timbers looked almost fresh. It seemed miraculous that such a tall, square, empty shell could have withstood more than a century of howling wind and driving snow without a scrap of steel in its wooden skeleton; the whole thing was held together with mortise-and-tenon joints and wooden pegs. I was taken by a sudden, irrational desire to own that barn, to become part of the story of a structure that had spanned multiple human lifetimes. I'm not sure why, but the moment I walked into that barn, I knew we had found our place.

GILLIAN AND I STARTED CONVINCING ourselves to buy the farm as soon as we got in the car. Over the years we have become expert in talking ourselves into things that normal people would think crazy, both of us having developed an intimate knowledge of each other's most irrational buttons, and how best to push them. Gillian and I are very different in many ways, but we both have the ability to drop everything and seize an opportunity. Unfortunately, we also have the ability to view buying a rundown farm in the middle of nowhere as an opportunity.

The main stumbling block was financial. We simply couldn't afford the $350,000 asking price. It took us a couple of days to think our way around this obstacle. We eventually came up with a two-pronged strategy: we would put in a lowball offer and we would move to the farm. That would allow us to rent out our apartment in Toronto, freeing up cash to make mortgage payments on the farm. It also gave us a double out: I didn't think Rubin would accept less than his asking price, so the offer was probably doomed to fail. And even if he did, we could sell the farm after a year if we didn't like it and move back to our house in the city. Vicki wrote up the offer and I promptly wrote off the whole endeavour, convinced that Rubin would say no.

A few days later, I was walking along Adelaide Street from the production offices where I worked to our studio at the CBC broadcast centre. It was a dark, wet Thursday evening. I was working on a weekly current affairs debate show called *CounterSpin*, which we described to our American guests as "like *Crossfire*, but less yelling." Ours was one of the few shows that still went live to air, and even after four years with the CBC it was a thrill to hear the theme music as the opening credits rolled and to be part of the choreographed chaos of the control room. I was keyed up, ready for the show, and had almost completely forgotten about our hare-brained attempt to buy a farm.

My phone rang. It was Gillian. "We got it," she said. "He accepted our offer." I stopped in the middle of the street, stunned.

As I went through the motions of taking the show to air that night, I started to feel that moving to the farm was meant to be. I had recently found myself looking around at the people sitting silently on the subway or walking hurriedly in and out of the office towers on Bay Street and asking myself, "What's the fucking point?" I had also taken stock of my life and realized that the times I was the most happy and fulfilled were the times when I was not sitting at a desk, when I was outside the city, when I was in the bush. I was always the "field guy" when I worked in Africa, the one who volunteered to take the five-day road trip though the jungles of Liberia a few months after the end of the civil war, to see what was going on outside the capital. I jumped at every opportunity to get out of the office and into the countryside. As a teenager, I had worked as a canoe-trip guide at a summer camp, sometimes spending weeks at a time in the wilderness.

It didn't matter where I was, I loved to get dirty, to sleep in shitholes, to interact with real people, to *do* things. But I had come to suppress that side of me. I fancied myself an intellectual, a sophisticate, an urban person. Smart people don't work with their hands. If you have a brain, being outside and performing physical labour are things you do on the weekends, or when you're young, not as a way to make a living. With the opportunity to buy a farm and live in the country suddenly before me, I began to question all that. I felt trapped in my desk job by the need to make money so I could visit the country on the weekends. Maybe I should just quit my job and live in the country all the time. Maybe deep down

I was a redneck who had accidentally been born into the body of a suburbanite. Maybe living on a farm was part of who I was always meant to be.

To my surprise, Gillian was even more enthusiastic about moving to the country than I was. She had spent much of her childhood resenting the work and isolation of living on a farm, but with us and the kids crammed into our second-floor apartment, she felt like a caged animal. She couldn't stand the fact that it took twenty minutes of organization to move both kids down into our tiny backyard, and that the weirdos who frequented our back alley made it necessary to move both kids back inside and upstairs if one of them needed a diaper change. Neither of us can remember the precise moment when it happened, but after a day or two of introspection, we had somehow decided to leave the city and move to the farm.

Vicki later told us that Rubin had been reluctant to accept our offer when she'd first presented it to him, but she'd sat with him at his kitchen table and talked it through. Rubin had received a similar offer a month or so earlier, from a Toronto lawyer looking for a weekend place. This guy had made it clear that the steel grain bin and manure pile in the backyard were both offensive eyesores that would have to be removed immediately as a condition of the sale, an attitude that rubbed Rubin the wrong way. (I learned later that Rubin was composting a dead cow in the manure pile, a detail that probably wouldn't have gone over well with the lawyer.) He eventually accepted our offer because his balance sheet included more than just the money we were offering. He looked at us as potential members of his community, people he would have to interact with on a daily basis. Our stated desire to live full-time on the farm, in close proximity to a big pile of shit, evidently counted for something. "You can't buy good neighbours," Vicki pointed out. Rubin agreed.

OUR NEWLY ACQUIRED FARMHOUSE NEEDED extensive renovations, so we decided to wait until spring to get started. In early June we hired a small moving van, loaded up all our possessions and drove north out of Toronto. We had rented a house in the tiny village of Dunedin, three miles down the hill from the farm. Dunedin was a thriving logging town in the mid-1800s, with two hotels, several churches and a school, but now it's just forty or so houses stretched along the Noisy River at the bottom of a deep valley that cuts into the escarpment. The only public space is the village hall, where eightieth birthdays and fiftieth anniversaries are celebrated and the community comes together for potluck dinners. Cellphone service didn't come to the village until 2014, when a new tower was built. Before then, the steep valley sides blocked all signals.

The place we rented was right next door to the only people we knew in the area. Steve McDonald and I grew up together. Our mothers taught at the same school before they were married, and his dad, Don, was my dad's best friend and the obstetrician who delivered me. Steve is a painter who had moved to Dunedin with his wife, Jackie, several years earlier, attracted by the stunning landscapes and the thriving local community of artists. Their second daughter had just been born when we arrived. Steve and Jackie introduced us to others in the village. Christine, a psychologist, and Kevin, a sculptor who made art out of old lab equipment and diagnostic machines he bought on the internet, lived on the other side of us. Tara, a naturopath, and Dan, a writer, lived down the road in a little house in a maple forest. They all had kids about the same age as ours.

We were immediately and enthusiastically welcomed into the community and started getting invited to social gatherings right away, although "invited" isn't really the right word. News of a social gathering—usually a bonfire in someone's backyard—would spread through the valley a few hours beforehand (or as people started to gather), and everyone would just show up. Potluck dinners would often occur spontaneously. I quickly realized that social norms were very different in Dunedin than they were in the city. No one I met ever asked me what I did for work, and no one seemed to think it was a big deal that neither Gillian nor I had a clear idea of how we would make a living on the farm. Almost everyone in the valley was self-employed. Whether they came from families that had lived in the area for five generations or had arrived just a year earlier, everyone lived in Dunedin because that's where they wanted to live; they had all cobbled together some sort of job for themselves after they arrived. Economists celebrate "labour mobility," the willingness of workers to move to where the jobs are, as a virtue of our modern economy, but for people in Dunedin it didn't work that way. For them, place came first.

THE SECOND OR THIRD DAY AFTER MOVING into our rented house, I got up early and headed up the hill to start renovating. It was a glorious, sunny morning as I drove up the steep, winding gravel road that runs due west from Dunedin all the way up the escarpment to the farm. Rubin's farm was the first on the left as I drove out of the village; he raised cattle. Then his brothers' and parents' places opened up on either side of the road; they raised sheep. The family also ran a small abattoir called McCormick Meats. Most people know our sideroad as "the McCormick Meats sideroad." The views over the McCormick pastures to the forest on the far side of the Noisy River valley are stunning. All the way up the escarpment I drove through forest and farm fields, until I reached the back of our place at the crest of the hill and saw the flat, green fields of my new domain stretched before me. It was a perfect day, and I was filled with what could only be described as joy.

Our farm is what's known as a "string hundred," a skinny block of land about three hundred and fifty yards wide and a mile long, with the long axis running east-west. The gravel road to Dunedin runs along the north side of the farm.

I turned down our long laneway and saw that the manure pile was still there. I also saw three or four pickups parked on the far side of the pile. I parked and walked over. Rubin and a bunch of guys I didn't know were leaning over the back of one of the pickups, having a chat. I hadn't seen Rubin in several months.

"How are ya now?" he asked as I approached.

"Fine, thanks," I said. The others just looked at me. I had started to learn that in our new community introductions weren't usually made right away. Some small talk was required first. "What's going on?" I asked, hoping to break the ice. The fact that six or seven guys I had never seen before were standing around in my backyard struck me as a little odd.

"Waitin' for the sprayer," Rubin replied. As I started to process this information, I heard the roar of an engine behind me. I turned to see an enormous self-propelled sprayer turn into our laneway from the road, on wheels that were far taller than I am. It drove straight past us without stopping, turned into the field closest to the house, the thirty-foot booms on either side unfolding as it went, and proceeded to disgorge a fine mist of herbicide over Rubin's newly planted corn. I sank to my knees in disbelief.

We had agreed to let Rubin farm our property for three years as part of the purchase agreement, and he had worked up part of the main field to the east of the house that spring and planted corn. Gillian and I were fairly committed organic consumers at that time. We had learned enough about conventional agriculture to know that it wasn't sustainable, and we didn't like the idea of feeding our kids pesticides. But the good food movement—that loose constellation of foodies and environmentalists, chefs and homesteaders—was still in its infancy, and there was a whole lot about conventional agriculture that we didn't know. We had decided that winter to ask Rubin to avoid spraying on our farm, but neither of us had gotten around to actually talking to him about it. So Rubin, like virtually every other conventional farmer in North America, had planted genetically modified, Roundup Ready corn, and he was now spraying it with Roundup to kill the weeds. The corn was about a foot high that day, but the weeds were almost as tall, creating a uniform green blanket across the field. Within a few hours everything except the corn would be dead and withered, and within a few days the corn would be the only living thing in a field of sterile brown dirt.

I didn't know it at the time, but an identical scene was playing out on farms all over North America that day: vast fields of corn and soy and canola were being doused with untold millions of litres of herbicide, gargantuan sprayers rolling across the land-scape. Gillian and I were profoundly naive to think that we could simply have asked Rubin not to spray. The entire system of agriculture that Rubin and just about every other farmer on the continent practised at that time was predicated on a complex regime of chemical pesticides and fertilizers. It still is. A conventional farmer can't farm without spraying. For conventional farmers, spraying *is* farming. Now I owned a piece of conventional farmland, and suddenly I was complicit.

The guys around the pickup just stared at me, not knowing why I was so upset. It took less than five minutes to spray the twenty-odd acres of corn that Rubin had planted, then the sprayer rumbled off to the next farm. I looked at my audience and decided it wasn't a good time to bring up the idea of not spraying next season.

THE ORIGINAL PART OF OUR FARMHOUSE was a white aluminum-sided box, about twenty-five feet square and two storeys tall. The ground floor was divided into two identical rectangles, one the kitchen, the other the dining room. The second floor had a central hallway and four more-or-less identical little bedrooms. A narrow, unfinished stairway led to an unfinished attic under the steeply sloping roof, with a single dormer facing north, toward the front of the house. A single-storey addition with a living room and the only bathroom in the house had been built onto the back in the 1970s. Most farms in our area didn't have indoor plumbing until the 1960s or even the 1970s, so the old toilet I pulled out of the bathroom may have been the first one ever installed in our house. The condition of the place was generally poor. On a visit late the previous fall, I'd found a snowdrift *inside* the attic, criss-crossed with numerous animal tracks. Most appeared to be from mice, but some I couldn't identify.

Demolition was the first order of business. I have always been reasonably handy and had performed numerous small do-it-yourself jobs on our house in Toronto with modest success, but demolition is really my forte. First I tore out all the shag carpet upstairs. Then my friend Alastair came down from Collingwood and we demolished the entire kitchen in one day. The linoleum came up easily, but underneath were several layers of thin plywood tacked down with little flat-headed nails on a four-inch grid. The plywood tended to splinter when I pulled it up, and the heads popped off the nails, leaving hundreds of tiny spikes sticking out of the original hardwood floor underneath. I spent several days on my hands and knees, pulling out rusty, headless nails with a pair of pliers.

Ella celebrated her first birthday and took her first steps in our rented house in Dunedin. Gillian came up to help out whenever she could but spent most of her time managing the kids; the farmhouse wasn't a great place for a one- and a three-year-old, what with the dust and the mouse shit and rusty nails sticking out of the floor and all. Rubin generously offered his big dump trailer to load all the debris into, and we quickly filled it up with old carpet, linoleum, plaster and wood. We had decided to convert one of the bedrooms into a bathroom, so I called a plumber who had an ad in the *Creemore Echo*, the local weekly.

Terry Nash arrived one morning to price the job. He was probably in his fifties, short and a little round, with a pure white beard. He bore a striking resemblance to Santa Claus, but he rode a Harley and used language that Santa would find profoundly offensive. Terry had spent most of his career in Toronto. He had moved to Creemore to semi-retire and found so much work that he got right back into the plumbing business. The wealthy weekenders and retirees who were buying up property in the surrounding hills were fuelling a building and renovation boom, so there was lots of work for the skilled trades.

Terry told me that even if I was going to do much of the carpentry myself, I would need to find an electrician, a well guy and someone to refinish the floors. When we got down to the basement, Terry stopped and looked around. "I remember this place now," he said. "The last time I was here, there was a calf down here." It turned out that when Rubin had lived in the place he sometimes kept sick calves in the basement, a dank, stone-walled space with an open stairway to the living room above. I was starting to understand that Rubin might be a little crazy.

Terry and his helper started breaking holes in walls and floors, extending pipes up to the second floor and to the attic, which would eventually become the master bedroom. One day while I was working away in the kitchen, I heard a shriek from the attic. I ran upstairs and met Terry's assistant, Dan, barrelling down in the opposite direction. He had been cutting a hole in the roof to extend the drain vent when he came across a three-foot-long snake coiled in a corner of the room. Gillian had brought the kids up that day, so she helped me direct the snake into a cardboard box and carry it out of the house. Rubin happened to be around, and he quickly identified our catch. "That's a milk snake," he said. "They'll coil around a cow's back leg while she sleeps and suck the milk right from her teat." Dan, Terry and the kids stared cautiously into the box, nodding in grave agreement. I checked on the internet that evening and learned that milk snakes hang out in barns not to drink milk but to catch the mice that live there. I wondered how many vermin must be living in our new house to support a three-foot-long snake in the attic.

The demolition eventually gave way to rebuilding, and our new and improved farmhouse began to take shape. The wood floors were sanded and refinished, revealing a forest's worth of species that had probably been cut on the farm when the house was originally built—pine, spruce, maple and ash. We bought new bathroom fixtures from IKEA, not realizing that Terry worked exclusively with North American brands. He spent many hours with his head in our vanity, shouting the vilest obscenities at Europeans in general and at Swedes in particular. I used a keg of beer to tempt my brother-in-law Neil (a skilled contractor) and two other friends (skilled beer drinkers) up from the city to help me install the kitchen cabinets. A predictable relationship developed between the amount of beer left in the keg and the quality of our work: as the afternoon wore on, the difficulty of drilling holes in the right places became ever greater, evidence of which is still visible in our kitchen. Rubin's dump trailer, which was full to overflowing by then, disappeared one day without warning. Rubin came by the next day, evidently very pleased with himself. "That whole load burned down to almost nothing!" he exclaimed.

He had taken all the renovation waste from our house, carpet, linoleum and all, and burned it at the back of his farm.

WHEN WE STARTED RENOVATING, we also started gardening. Gillian's parents had grown up in a small town in western Illinois near the Mississippi River, married young and moved out to Vermont to homestead. They had always worked off-farm but kept a huge garden, as well as raising laying hens, meat chickens and sheep. They even had a cow when Gillian was young, which her father would milk by hand twice a day. One of Gillian's earliest memories is of sitting down on the front porch to a summer meal sourced entirely from the farm. I had tended a vegetable garden in our suburban backyard while growing up and I'd had a decent strawberry patch for a number of years, but my gardening chops were nothing like Gillian's.

Our farm had belonged to Rubin's Aunt Marge and Uncle Russell Rowbotham in the 1970s. Aunt Marge's original vegetable garden was still there when we moved in, between the house and the barn, surrounded by an old split-rail fence that had remnants of chicken wire tacked to it to keep out the groundhogs and rabbits. It probably hadn't grown anything but grass in over fifteen years, but Rubin had kept calves in it the previous summer that had done the dual service of eating all the weeds and depositing a generous amount of manure. Digging into that soil was like slicing a Black Forest cake; the earth was rich and light and black, and full of worms. It smelled beautiful.

We planted all the regular stuff—green beans, head lettuce, beets, carrots, zucchini, a few potatoes and beefsteak tomatoes. We didn't do a particularly good job of spacing things or of weeding, but in soil like that it didn't matter. Everything flourished.

As the renovations stretched into the summer, I learned more and more about our old farmhouse and the community we were joining. One of my best sources of information was Lorne Bunn, the electrician who was rewiring the house. Lorne didn't do any of the work himself—he had two young apprentices who pulled the wire and hooked up the new lights. Lorne priced the jobs, collected the cheques and shot the shit with the clients. It didn't matter how busy he purported to be or how hard his employees were working, Lorne always had time to tell a story or two. Lorne is solidly built, with close-cropped hair and a big round face. He always wears steel-toed boots laced up over his jeans. He was in his sixties when we met him, born and raised in the area, with an encyclopedic knowledge of local genealogy and history. He had a perpetual half-smile on his face, as if he was always about to tell you a secret.

Rubin had already told me about the previous owners of the farm ("Eye-talians in the lumber business"), who had used the place only on weekends. Rubin's Aunt Marge and Uncle Rusty had owned it before them, farming it in the 1970s and 1980s before moving into Stayner, about twenty minutes northeast of the farm. Lorne had known the Fachnies, who lived there in the 1960s. Most old-timers still refer to our farm as "the Fachnie place."

I was never quite clear on how many Fachnie kids there were, but it seemed like quite a lot. "One of the kids wasn't quite right," Lorne told me, which I understood to mean she had some sort of mental disability. "They used to throw big parties, but whenever company was over, they would lock her away up in the attic. The only time I ever saw her was one day when we were leaving, and I looked up and she was standing in the little window in the roof, all dressed up, waving at us. It seems terrible now, but that's the way you dealt with those kinds of things back then." A chill went down the back of my neck, thinking of that girl in her party dress, all alone in our cold attic while everyone danced and caroused below. I don't know what became of her, or the rest of the family. Gillian can be a little superstitious, and I didn't know what she would think of the girl in our attic. I didn't tell her about it for years.

WE FINALLY MOVED INTO THE FARMHOUSE in August, more than a month behind schedule. The attic still wasn't finished, so Gillian and I slept on a futon in the tiny second-floor office, while our king-sized bed was used by the kids as a trampoline in the dining room. We didn't have much furniture yet, but the house was now bright and airy, and it felt like home. We had installed a big new window in the kitchen above the sink, and the view out over our fields to the east as we prepared dinner each evening was spectacular. The hot weather and huge front lawn inspired our kids to remove their pants at every opportunity, and they spent many hours running around on the grass bottomless, or sometimes stark naked.

Despite an almost total lack of attention, our garden began to produce prolifically. The original garden was only about twenty feet square, and we had planted less than half of it. But the amount of produce that came out of that little area was astonishing. We ate huge salads with every meal. Gillian made all the standards that her mom and dad had eaten growing up in the Midwest—roasted beets, snap peas with cottage cheese, maple-glazed carrots, and mashed potatoes with enough butter to make a cardiologist blush. For me, the flavour of the stuff coming out of our garden was a revelation. It had been many years since I had eaten vegetables grown in beautiful soil, fresh out of the garden. Everything tasted amazing. Our perennial battles to get our kids to eat their veggies suddenly stopped. They both ate everything we put in front of them, without discussion. We built a long makeshift harvest table out of old barnboard we had found stacked in the attic and recreated Gillian's childhood tradition of enumerating all the ingredients that had come from the farm at the start of each meal.

As summer slipped into fall, I was still spending most of my time working on the house, but the tradespeople had all gone and I was forced to start thinking about obtaining gainful employment. The last outside contractor was the fireplace guy, who installed a large steel wood stove in the living room in the middle of October. The weather had turned at the start of the month, and we had woken up to snow on the ground more than once already. The fireplace guy showed me how to light the stove and told me that if I stoked it up before I went to bed there would still be a healthy bed of embers in the morning to get the fire started again.

He was right. I used one match to light the fire that day, and that fire burned continuously for the next five months. All winter, through brutal cold and blinding snow, I stuffed cord after cord of wood into the stove and sunk deeper and deeper into unemployment and isolation. I quickly understood why the old-timers I had met looked at me a little funny when I told them where we were living. "Windy spot, up there," they would say, with an air of warning. I lay in bed at night listening to the wind roar and the house creak, wondering what we had done.

CHAPTER 2

BABY STEPS

I DID A LOT OF THINKING that winter. Gillian had made a deal with her old consulting firm to work as a freelancer and had picked up several contracts in the city. You could drive from our farm to downtown Toronto in about an hour and a half on a day without traffic or bad weather, if such a day were ever to occur. It never did. Gillian often spent three hours or more battling traffic and blowing snow to get to work, and would sometimes arrive home late at night completely traumatized by the harrowing drive. She started spending a night or two in the city every week to cut down on her commuting, leaving me alone with the kids for long stretches. The weather at the farm verged on apocalyptic most days, so the kids couldn't spend more than a few minutes outside. The wind blew so hard at times that the water would slosh around in the toilets. I felt like we were living on a boat that was caught in a perpetual storm. After a couple of weeks of staring at each other by the fire, the kids and I agreed that they needed some other form of socialization. I signed them up at the daycare in the village of Duntroon, about fifteen minutes north of the farm.

With all of my renovation tasks completed, the kids in daycare and Gillian at work, I suddenly realized that I was unemployed. I spent long hours alone in the farmhouse, stoking the fire and racking my brain, trying to think of what I should do with my life. In the past I had always done a good job of seizing opportunities, but I had never really identified a grand plan, my mission in life. Something would always come up and I would grab it, then ride it until I got bored and something else came along. But nothing seemed likely to present itself while I sat alone on the farm, buried in snow.

I thought up and promptly rejected a long list of purely mercantile ideas: making furniture, selling adult-sized footed pyjamas on the internet. I decided for a while that I would be a writer, and even published a couple of magazine articles, but I felt too isolated to write about anything with conviction, and the pay was terrible. I filled my time reading very fat books that I had been meaning to read for ages: a chronicle of the Second World War, a biography of Che Guevara, *War and Peace*. I found myself drawn to books about people who were wholly committed to a clear and momentous cause, fully engaged in their moment in history. But reading about those people made me feel like a loser. I had been a successful journalist in Toronto and had worked with democracy and human rights activists all over the world before that; I was engaged in more than one momentous cause when I worked overseas or in the city, but I had thrown all that away on a whim and moved to a farm. I felt like I had picked the worst possible place to find my true calling. I felt like I had blown it.

I was also reading a lot about climate change. In February 2005, as I lay on the couch, trying to conjure up my own future, the earth experienced its 240th consecutive month of above-average temperature. That trend has continued to this day— thirty-one years and counting—with every single month above the twentieth-century average. Every year the data becomes more robust, the climate models more sophisticated and the evidence more compelling, and every year we realize that our most dire predictions probably aren't dire enough. Even in 2005 it was difficult for a reasonable person to look at the available evidence and not conclude that humanity is pretty much fucked, that climate change is probably the biggest challenge, not just of our generation but of any generation. The philosophical exercise of figuring out my own future became complicated by the realization that the future of our entire species is in doubt.

That spring, Gillian gently suggested that the solution to my existential crisis might become evident if I got off my ass and went to work. Gillian is a fiercely practical person who would chew glass if necessary to provide for her family, but she was not willing to go on commuting to the city indefinitely while I stayed in my pyjamas all day, searching for the meaning of life. So when my new friend Jamie offered me a job helping him renovate his house, Gillian gladly accepted.

JAMIE LIVES IN LAVENDER, A HAMLET of five or six houses on top of the escarpment on the opposite side of the Noisy River valley, just over five miles southeast of our farm. He was born in Toronto a couple of years after I was and also travelled the world before living for a time in a crappy cabin on some family land in Mulmur, the next township to the south. He and his wife, Juliette, had met while living on Haida Gwaii, on the west coast, and had bought a rundown old house at the four corners in Lavender a couple of years before we moved up. Jamie has never been a big fan of deodorant, and he has a head of hair that makes Einstein look slick. He didn't belong in the city. He started working for a local builder when he first moved to the area, just to make ends meet, and found that he had a natural talent for it. We had hired him the previous fall to finish the attic conversion at our place.

I spent several weeks helping Jamie tear the roof off his house and build a big shed dormer in his master bedroom. He paid me fifteen bucks an hour, cash, which was about the going rate for semi-skilled manual labour back then. At that rate, my contribution to the family finances was largely symbolic, but at least it got me out of the house. Working for Jamie turned out to be fortuitous for another reason: he was an avid gardener. Jamie and his friend Mark, a high-school drama teacher who lived in Avening, south of Creemore, had both planted large vegetable gardens in their backyards. The previous summer they had started selling their surplus produce at the Creemore Farmers' Market. It was just a hobby, but they were kind of obsessed with growing food. They bought a big Rototiller that they hauled back and forth between their gardens, and they talked incessantly about vegetable varieties and growing techniques.

When Jamie's job was finished, I picked up a similar position with a small contractor near Blue Mountain. I can't even remember the guy's name, but I do remember that he ran an extremely shoddy operation. There were three or four of us working for him on a farmhouse reno job. All the other workers were there on some kind of juvenile work-release program—I was the only member of the crew who had never been convicted of stealing a snowmobile.

When that job ended, I wound up working for a friend of a friend who owned a painting company. I was assigned to work with the company veteran, a guy named Patrick Keating, commonly known as Travelling Pat on account of his winter excursions to the far corners of the globe. Patrick is the same age as me, but he's an old soul and he knows pretty much everyone for miles around. He lives in a converted auto mechanic's shop at the top of the Mad River gorge, just east of Singhampton. I once went over to see Patrick and ended up sitting at the kitchen table talking to Ayrlie, his partner, for half an hour while she had a bath. Their tub is in the middle of the kitchen.

Our first job was to repaint the window and door frames on an old brick building in Creemore that was once a bank and was being converted into a café. Creemore is sort of the administrative centre of our lives, about six miles straight east of the farm. It's where our kids went to primary school and where Foster played hockey. There's a small grocery store, a bank, a post office, a hardware store, a pub and a couple of restaurants. Creemore is our postal address. It's the closest town that people from outside the area might have heard of; it's where we say we're from.

The main drag in Creemore is Mill Street. The north end is lined with huge maple trees and grand old houses. The south end is the downtown, with beautiful two-storey, flat-fronted brick buildings built right against the sidewalk, typical of small-town Ontario. What makes Creemore unique is that the main street doesn't go anywhere. It just sort of peters out at the south end by the Mad River. You don't pass through Creemore to get anywhere else, so the whole town feels a little bit tucked away, self-contained and autonomous.

Two businesses define the village: the brewery and the newspaper. Creemore Springs was one of the first microbrewers in Ontario, and it's still the largest employer in town. It's housed in an old brick hardware store at the south end of Mill Street, and tanker trucks rumble back and forth to the spring just west of the village that provides the water for the beer. The *Creemore Echo* ensures that everyone knows what everyone else is up to. Hilarious multi-week arguments between readers can develop in the letters to the editor section, which send everyone in town to their mailboxes on Friday mornings, eager for the latest installment.

Together, the brewery and the *Echo* provide the economic and social heart of Creemore. A tight-knit community, with lots of beer—what could be better?

IT WAS A GLORIOUS, SUNNY MORNING in early June when Patrick and I started painting the old bank. We had to set up our ladders right on the sidewalk of Mill Street to get to the second-storey window frames. Pretty much as soon as we started, people began stopping to see what we were up to. I don't think a single person walked by without at least acknowledging our presence. At a minimum, they would say hello or make a joke about our painting ("You missed a spot!"). Most would stop and examine our work, and many would engage in prolonged conversation. What surprised me was that Patrick would stop working completely whenever someone appeared, come down off his ladder and have a relaxed, unhurried chat, sometimes for a full five or ten minutes. He didn't seem the least bit worried that it took us the whole morning to finish a couple of window frames that should have taken half an hour. Most of the passersby knew Patrick, or at least knew of him, but I was obviously new in town. Many people would ask Patrick questions about me, even though I was up a ladder just a few feet from where they stood.

"Who's the new guy?" an elderly woman asked.

"Oh, that's Brent," Patrick replied cheerily. "He moved into the old Fachnie place, up past McCormick Meats."

"Windy spot, up there!" she observed, barely looking up at me.

I smiled and nodded from up on the ladder. I half expected Patrick to convince the old woman to take over painting for him so he could go take a nap.

The nadir of my underemployment was reached later that month, when I spent an entire week staining deck chairs at the weekend home of a wealthy Toronto banker. He had dozens of those slatted wood Adirondack chairs. It was horribly tedious work. I spent the whole week covered in red stain, feeling sorry for myself and lamenting the fact that I was pissing away my many years of education and experience for fifteen bucks an hour. I had worked with heads of state and Nobel laureates on four continents, confronted child soldiers on the streets of Liberia, and hung out with African National Congress activists in a shanty in Soweto. Now I was staining some rich guy's deck chairs. The worst part? I wasn't even very good at it. I started to wonder if we should throw in the towel and move back to the city. The problem was, I loved our new community. I loved the people we had met. I loved the landscape we were living in. When I wasn't worrying about my future or how I would make money, I simply loved living on the farm. For the first time in my life, I felt like I was home.

MY PAINTING CAREER WAS MERCIFULLY interrupted by a long holiday weekend in July. Gillian and I had been invited to spend it with friends in the Laurentians, north of Montreal. I had worked with Barb at the CBC, and her husband, Patrick, universally known as Patty, had a family place near the village of Saint-Sauveur with several hundred acres of forested hills, a substantial lake and a little old cottage that was crammed with enough beds to sleep two dozen people. We loaded the kids into the car and drove all day to get there.

The morning after we arrived, Barb suggested we drive to a local farm to buy some food for the weekend. We had a minivan at the time (another symptom of my fall from globe-trotting coolness), so Barb, Patty and their two boys piled in with us. I was reluctant to go, having spent the entire previous day in the car, and the forty-five-minute trip on winding country roads made me grumpy, but what we saw that day would completely change our lives.

Runaway Creek Farm was nestled in a broad valley surrounded by forested hills, with a small creek meandering through fields that seemed impossibly lush and green. The barns and buildings were all freshly painted and in perfect repair, and the whole place was orderly and well kept. Runaway Creek was a classic diversified organic farm, the kind of place that is now widely celebrated by chefs and foodies. But back then, to Gillian and me, it was completely novel. There were vegetable gardens, a large flock of hens, sheep and cattle and a bunch of pigs. The owner was there. He showed us around and told us all about his operation. He sold his produce at his farm stand, at the farmers' market in Saint-Sauveur and to the many local restaurants throughout the Laurentians that served the skiers and summer tourists. We couldn't believe that such a place existed. It was as if the aesthetic ideal of our new farm, with animals and gardens and beautiful views, had been transformed into a tidy and bustling business.

The kids ran around happily, trying to catch the hens and poking the pigs through the rails of the fence around their pen. Barb and Patty loaded up on eggs, pork chops and copious amounts of beautiful vegetables. Gillian and I walked all over the farm, just taking it all in. The wheels began to turn. I had been flirting with the idea of farming since we bought the farm, but the kind of farming that Rubin and all our neighbours were doing—big machinery that cost a lot of money, growing corn and soy and canola—was totally unappealing. I hadn't dared tell Gillian; her unhappy memories of farming centred on being dragged out of bed at five o'clock in the morning in February to feed the lambs before heading off to school. Her parents had real jobs off-farm, so for Gillian, homesteading had seemed like a lot of pointless work. But Runaway Creek was somewhere between those extremes. It was a business, but with a style and scale that seemed both manageable and profoundly fulfilling.

I had made several wrong turns on the way to the farm, so Patty took the wheel on the way back and Gillian and I sat huddled in the back of the van. I was excited, but also worried that Gillian might not be sharing my thoughts. She hated commuting to Toronto, but Gillian was also ambitious and was rapidly climbing the corporate ladder; her firm had recently offered her a partnership. My most recent work experience had been painting deck chairs—I was ready to jump on just about any alternative. But Gillian had a lucrative, prestigious career path before her.

"That's what I think we should do!" she blurted out as we started driving. A wave of happiness and relief washed over me. "Let's do exactly what that guy is doing!" I almost shouted. It instantly made sense. Here was an opportunity to create a business that expressed our values and ideals, that allowed us to stay on the farm and still make money. Gillian in the end didn't care about the partnership or the salary or the trappings of her Toronto consulting job. She wanted to do something real and to spend time with her children. We had spent more than a year in limbo, wondering how to shape our family's future, so when we saw that farm, we grabbed hold in a way that probably wasn't entirely rational.

Barb and Patty got caught up in our excitement, so when we got back to the cottage, we dug out a bottle of champagne that had been left in a cupboard after some forgotten New Year's party and celebrated the birth of our new enterprise. We hadn't thought to ask the farmer even the most rudimentary questions about the viability of his business. We didn't even know his name. But that didn't matter. We had a plan.

A few days later, on the long drive home, Gillian and I fleshed out our strategy. We would start expanding our garden right away and would launch our farm business the following spring. We would use the rest of the summer and the winter to read, ask advice, buy equipment and prepare a business plan. We would start small.

Looking back now, it seems idiotic that it took us a year on the farm to come up with the brilliant strategy of—wait for it—starting a farm. But things were very different in 2005 than they are now. *The Omnivore's Dilemma*, by Michael Pollan, was still a year away from publication. Mark Bittman was writing cookbooks, not pointed criticism of the food system. Eric Schlosser's *Fast Food Nation* had come out a few years earlier, and I had read it, so I had a basic understanding of the ways in which the industrial food system was broken. But I didn't yet understand that there was an alternative. What made our visit to Runaway Creek Farm so transformative is that we got to see the alternative fully formed, in the flesh. It was as if our future had been made manifest and laid before us.

The more we talked about farming, the more I convinced myself that this was the grand project that I had been looking for. This idea, that we were embarking on some kind of mission, changed everything. Looking back now, more than a decade later, it seems to me that the particular circumstances of the launch of our farm made everything that happened afterwards, both good and bad, almost inevitable. What were those circumstances? There were many, but the most important were these.

First, we started with a total lack of knowledge and experience. This set us up for innumerable failures, but it may also have led to our ultimate success. We didn't come to farming burdened with preconceived notions or received wisdom. In fact, we came to farming with no wisdom whatsoever. We were adherents to neither the chemical-industrial model of conventional farming nor the idea that growing food was some sort of pagan spiritual endeavour. We didn't have the money or inclination to buy big tractors and to grow commodities at scale, but we also didn't think that planting more than a quarter-acre would be selling out. Gillian retained some knowledge of the mechanics of growing vegetables and raising animals, but her parents' farm had never really been a business, and she had left almost twenty years prior. Gillian's emotions when looking back on her childhood on the farm were decidedly mixed, and neither of us was motivated by some sort of 1970s-style back-to-the-land utopian vision. We had never heard of Wendell Berry. We didn't know what CSA stood for. We entered farming in a state of intellectual lightness, open to all possibilities. Our ignorance set us free.

Second, we needed to make money. Gillian and I were equally caught up in the idea of creating a farm, and neither of us is the kind of person who is happy to take the back seat, so we decided at the very outset that the farm would have to support both of us. We didn't know that the average Canadian farm loses money, that the large majority of farms are kept afloat only by income earned off-farm by one or both spouses. It didn't seem like a lofty goal at the time, but aspiring to create a farm that was profitable enough to support our family was aiming very high. So, with ignorance again on our side, we set out with the clear idea that our farm would be a business, and a profitable one at that. In this respect, we had no fallback position. I was effectively unemployed. Gillian hated her job, hated commuting and hated being away from the kids so much. She was done. There was no plan B.

Third, we wanted to do something that benefited our community. The food that we had produced in our little garden on the farm was so delicious and so beautiful that we had a strong desire to make that food available to our friends and neighbours. Gillian especially was dismayed that it was almost impossible to find healthy, organic food around Dunedin and Creemore—the selection in our local grocery stores and markets was far worse than it had been in downtown Toronto, and Ontario was light years behind her home state of Vermont, where organic farming was flourishing like nowhere else on the planet. So making good food accessible to everyone was a primary motivation. Our farm would feed and nurture not only our family but also our community.

Finally, we had to do something about climate change. My winter of introspection had convinced me that now was the time to stop worrying and to start doing something to avert an environmental catastrophe, and that my grandchildren would scorn me on my deathbed if I didn't act on my convictions. I was just beginning to understand the colossal contribution that agriculture makes to the problem of global warming. But even with my limited knowledge, I knew that changing the food system was a project grand enough to assuage my guilt that I had, until then, been fiddling while Rome burned. So the *way* we farmed would be just as important as making money. We set out to create a farm that would be profitable *and* sustainable, one that would be a model for broader change in our food system. From the outset, farming for us would be an overtly political act: we wanted to prove that it could be done. We started with chips on our shoulders, chips that grew dangerously large over time.

Thus began our farm dream, with a clear financial goal (profitability), an overriding environmental and political imperative (addressing climate change), a social mission (accessibility), and two protagonists who knew almost nothing (embarrassing, but self-explanatory). The die was cast. It was decided. We would farm.

THE MOST IMMEDIATE BENEFIT of our decision to start a farm business was the excuse it provided to quit my painting job. I was now devoted full-time to the task of preparing for next year, our inaugural growing season. Gillian had scaled back her consulting work for the summer but still had a project or two on the go. It was amazing to see how she could transform herself from farm worker to professional consultant in the blink of an eye. We would be out in the garden pulling weeds, sweaty and covered in dirt, when Gillian's phone would ring and she would instantly switch into city mode. Everything about her demeanour and tone of voice conveyed the image of someone sitting at a desk in a fancy office somewhere in the city, when she was actually up to her ass in tomato plants. I'm sure none of her clients was ever the wiser.

We doubled the size of the garden that year and took on the ill-fated chicken experiment that ended with the bludgeoning of five unfortunate chicks. That first morning in the coop, as horrible as it was, helped me approach farming in a much more realistic and humble state of mind. The rose-coloured glasses were off. I now felt acutely the responsibility of caring for our forty-five remaining chickens. I didn't want to again subject them to undue suffering or harm. But the coop held, there were no more varmint mishaps and the birds flourished. The Special Dual Purpose chickens were all healthy, active and happy-looking, though they were pretty leggy, kind of like white versions of the Road Runner.

Our most pressing task that summer was opening up new land for growing vegetables. Gillian and I decided on a small plot inside the big hedge north of the house. This was technically part of our front yard, but it was well sheltered from the wind and on the opposite side of the laneway from the house. It was less than a fifth of an acre, but it was four or five times bigger than the current garden, which seemed like all we could manage. We paid a neighbour who had a small tractor to plough under that part of our lawn.

Gillian and I knew enough about the fundamentals of gardening to know that we would have to do more than just kill the grass

47

to create a productive vegetable garden. Luckily, Rubin's huge manure pile was still in the backyard (the heat of the composting pile had reduced the cow to a few tiny shards of bone). I think he felt guilty that he had been promising to move it for a year and a half, so he said we could use as much of it as we wanted. He even offered to lend me a small tractor to move the manure. Farmland in Ontario is taxed at one-quarter of the usual rate, as long as it's being worked by a registered farmer. Rubin managed a number of farms owned by weekenders, meaning he kept cattle on them or cut hay in exchange for signing their tax forms. One Wednesday, when we knew all the weekenders would be busy at their desks down in the city, Rubin took me over to one of "his" farms on the other side of the valley and we liberated a beautiful little tractor with a nice bucket on the front. "Just return it in a week or so," he told me. "The owner'll never know it's gone." I felt pretty pleased with myself, loading manure onto our new garden with a tractor expropriated from an absentee landowner. Che would have approved.

THAT AUGUST, WE ONCE AGAIN STARTED enjoying the full bounty of our garden. We had planted a lot more tomatoes and had branched out from the generic beefsteaks we had grown the year before, experimenting with heirloom varieties for the first time. The term "heirloom" is thrown about pretty casually these days, and there's no agreement on a standard definition. But it usually means that the variety has been around for a long time (though not always), it was created through traditional plant-breeding techniques (rather than hybridization or genetic manipulation) and it will breed true, meaning that if you plant a seed from an heirloom tomato, the plant that grows will produce tomatoes that are the same as the one you got the seed from. We planted Brandywines, San Marzanos and a variety called Copia that produced huge, gorgeous fruit shot through with streaks of green, red and fluorescent orange. My favourite was a variety called Yellow Mortgage Lifter—some of the tomatoes we got off those plants weighed two pounds apiece. My standard lunch was a sandwich made with a single thick slice of yellow tomato on bread that Gillian had baked, with lots of salt.

Then came September, and time to kill the chickens. The Special Dual Purposes had lived up to their billing as excellent free-rangers. We let them out of their coop every morning and they roamed all over, around the barn and the house, eating grass and bugs and scratching in the dirt. We had an infestation of tent caterpillars in our poplar trees that summer, which turned into white moths in August. One day a big wind suddenly came up, blowing hundreds of moths out of the trees. The whole flock of chickens came charging over, running in all directions, grabbing moths out of the air as they floated down. In the evening the chickens would wander back to the safety of their coop and we would simply close the door before it got fully dark. We lost only one bird that summer after the initial cat incident. I looked out one morning to see the flock pecking about in the yard, then looked back a few seconds later to see all the birds gone, and white feathers scattered across the lawn. I think a large hawk or an owl must have taken one and the others had hightailed it back to the barn.

The kids were two and four that summer, and they grew particularly fond of the chickens. They would herd them around, try to catch them and occasionally pick one up and carry it around, incorporating the birds into whatever game they were playing in the sandbox we had built. As the time came to kill the birds, Gillian and I worried about how the kids would react. Gillian had spent more than a decade as a vegetarian after witnessing a pig kill on the farm as a girl. We didn't want to traumatize the kids by killing all their playmates. So we did what parents have always done in such situations: we lied. I told Foster and Ella that it was getting toward fall, and one day soon we would wake up to find that all the chickens had flown south for the winter. I told them not to be sad—chickens migrate. It's nature's way.

We booked in the chickens at a small poultry-processing plant near Dundalk, a thirty-minute drive to the west. On the scheduled morning, Gillian and I got up very early and snuck outside so we wouldn't wake the kids. I had borrowed chicken crates from a neighbour in Dunedin. Chicken crates are shallow, rectangular, slatted plastic boxes that prevent the chickens from climbing on top of and smothering each other during transport. You can fit about six meat birds or ten laying hens in one crate.

Chicken catching is hard and dirty work. Anyone who grew up in rural North America before the 1980s will have a chicken-catching story—it was once common summer work for teenagers. Now it's done almost entirely by migrant labourers. It's usually easier with meat chickens than laying hens, because the meat birds are so much bigger and slower, but our Special Dual Purpose birds were fast and agile. Gillian and I didn't know that the standard practice is to catch the birds at night, when they are half asleep and dopey. We went in at the crack of dawn, when the whole flock was very much awake and ready to charge out for a day of free ranging.

We decided that Gillian would work the crates, opening the trapdoor lid, stuffing in the birds and ensuring that the birds already in the crate didn't escape. I would do the catching. The first bird was easy. They didn't know what was going on, so I just walked up and grabbed one by the legs. But then it got harder. The more we caught, the more the remaining birds freaked out. Chickens, you might be surprised to learn, can fly quite well over short distances. As the level of anxiety in the coop rose, so, too, did the altitude the birds could reach. Soon I was pursuing birds in three dimensions, all around the floor of the coop and up to the level of my head. The jumping, running, flapping birds filled the air with straw and feathers and flying flecks of shit. It was a cool morning and I was wearing canvas overalls and a heavy flannel shirt, but soon I was sweating profusely and had to take off a layer. When I did catch a bird, it would try to peck my hands and flap so hard that its wings would bruise my forearms. If I didn't hand it to Gillian very carefully, she would get a face full of flapping wings as she tried to get the bird into the crate.

The first bird I caught was unaware of what was going on, hemmed in by the rest of the flock, and maybe a little slower and less agile than average. When it was removed, it left a little more room for the others to manoeuvre, and its capture let the others know that something was up. Each bird that I captured meant more space, more anxiety and more adrenaline (or whatever the equivalent chicken hormone is called) for the remaining birds. Each bird I caught was more difficult than the last. I'm sure a mathematician could make a beautiful graph of this phenomenon, plotting effort against number of birds remaining to be caught and coming up with some sort of exponential curve.

By the time we were down to four or five birds, it was almost impossible. I was diving, sprawling and panting, covered in shit and sweat and bits of straw. Gillian and I had to double-team the last one, a small, incredibly fast female that was rocketing around the coop like a pinball. It took us well over an hour to catch our forty-four birds. We were exhausted.

As we loaded the crates into a borrowed pickup, I heard a sound behind me. I turned around to see Foster standing right beside the truck. He had a Spider-Man pyjama top on but was naked from the waist down. He was holding an empty toilet paper tube.

"There's no paper left," he said. We stood staring at each other. Foster didn't seem to register that his parents were in the midst of loading all the chickens onto a truck, both covered in shit and feathers. I felt like the Grinch when he's confronted by Cindy-Lou Who as he shoves the Christmas tree up the chimney. But I didn't need to make up a lie. Foster didn't ask what we were doing; he was intent on getting his ass wiped. "Just go in the house and I'll be there in a second to help you," Gillian said. He turned around and walked away.

THE NEXT DAY WE ROASTED OUR FIRST Special Dual Purpose. They didn't look like supermarket chickens when we got them back from the slaughterhouse. They were pretty scrawny. The first one's legs stuck up so high that we had to remove all the racks to fit it in the oven. It was tough and lean and difficult to carve, but the flavour was incredible. It reminded me of the village chickens we ate in Africa. I gave one to Ivan at Hamilton Brothers, as promised, though I was a little embarrassed. "That was a funny-looking chicken!" he said with a huge grin when I asked him later how it was. We eventually learned that roasting isn't the best way to cook a Special Dual Purpose; they are more of a soup or braising bird.

Foster never mentioned our encounter at the chicken coop that morning, and the migration ruse worked for a while. He had started kindergarten in Creemore a few weeks earlier, the bus picking him up at the end of our lane for the first time one sunny September morning for the fifteen-minute run into town. The school was a beautiful two-storey brick structure built in 1916, with broad wooden stairs that were deeply worn by generations of local kids running up and down. On the first day of school, the janitor climbed up into the old bell tower and rang the original brass bell, which could be heard all over town. Foster was in a class that was more than half farm kids, with all the old families represented: Carrutherses and Millsaps and Taylors.

One day Foster arrived home on the bus with a grave look on his face. "Dad, where did the chickens go?" he asked.

"I told you, they migrated," I said. "They flew south for the winter."

"I think you're joking," he said, with an accusatory tone (back then, Foster used "joking" as a synonym for "lying"). "Maya said that chickens don't migrate." Maya was Kevin and Christine's daughter, down in Dunedin. She was a year older than Foster, and far too smart for her age. "Dad, are the chickens in the freezer?"

"Yes," I confessed. I felt terrible.

"Okay," he said, and skipped off. He continued to eat chicken, with gusto.

WINTER CAME EARLY AGAIN THAT YEAR, but this time we were prepared. In the early fall I had bought a huge old-school secondhand Husqvarna chain saw from Hennessy's in Duntroon that I grew to love (it made me look like a total badass) and to hate (it was unbelievably heavy). I used it to cut up all the big branches that had been left in our maple bush after Rubin logged the forest two years before. I cut, split and stacked four bush cords of firewood before the first snow flew. We had a freezer full of chickens, tomato sauce and spinach, as well as carrots, onions, garlic, winter squash and potatoes hanging from the rafters or stacked in crates in the basement. We were fully provisioned for a winter of reading, planning and strategizing. In the spring, we would farm.

CHAPTER 3

SCHOOL OF HARD KNOCKS

THAT WINTER, I plumbed the depths of my farming ignorance and found them to be very deep indeed. What I didn't know about growing food could not just fill a book, it could fill shelf upon shelf of books, a whole library of facts and information that I did not yet possess. But slowly, through many hours of reading and internet searching, I began to fill up that empty well of knowledge. Gillian was starting from a slightly better place than me, but she, too, read voraciously. We sometimes fought over a book that both of us were trying to read at the same time, or traded back and forth between chapters.

We quickly realized that there was a surprisingly strong consensus about how to run a small, sustainable, diversified farm. Many of the books we read had similar recommendations when it came to farm scale, crop and livestock mix, growing techniques and marketing. When we talked to organic farmers and asked advice from people who were already doing it, we got the same story. Gillian and I dubbed this the "small farm orthodoxy," and it went something like this: stay very small (five or maybe ten

acres, max), grow as many different kinds of vegetables as possible (with a strong emphasis on heirloom varieties), keep some
livestock around (chickens and hens and pigs for sure, and maybe
sheep and cattle if you're feeling ambitious), and sell what you
produce directly to consumers. The selling part was probably
the most prescriptive: everyone seemed to assume that all small
farms would run a Community Supported Agriculture program,
or CSA, where customers pay up front for a weekly delivery of
whatever the farm is producing throughout the season. Selling in
a farmers' market was also a given, and maybe running a farm
stand. A few people talked about selling directly to restaurants.
Wholesalers, supermarkets and middlemen of any description
were evil and to be avoided at all cost.

We also realized quickly that of all the books and articles on
how to start a small organic farm, one towered above the rest:
The New Organic Grower by Eliot Coleman, the granddaddy of
the entire genre, published in 1989. Coleman had run Four
Season Farm in Maine for many years. He had written a number of how-to books, and even had a TV show at one point. *The
New Organic Grower* was incredibly detailed and prescriptive; it
quickly became our bible. Coleman talked about everything—
how to buy land, what tools and machinery to use, how to hire
workers, how to plant, weed, harvest and sell vegetables. He even
explained, with illustrations, the proper way to pick a tomato. To
be fair, Coleman stressed repeatedly that all farms are different,
that every grower must develop their own methods, and that his
way of doing things was only a suggestion or starting point. But
his formula was far too seductive for us to resist. We knew nothing. Eliot knew everything. We decided to do exactly what Eliot
Coleman told us to do.

Plans for the farm were progressing well, but by mid-winter we
were confronted with a serious problem: we didn't have enough

money to buy all the stuff we needed. Gillian had reluctantly gone back to her consulting work in the city, which allowed us to cover the mortgage and pay the bills, but we didn't have anything in the way of savings to invest in the business. I was in the midst of an intensive "start your own business" course that was offered free to the unemployed, which I technically was, since I was paid under the table for all my renovation and painting work. The course was taught by a ridiculously enthusiastic serial entrepreneur named Cliff, who was well into his seventies but had the energy of a twenty-year-old. We met in a small, windowless classroom in Owen Sound for a full day, once a week, for ten weeks. The course wasn't geared toward farming, and that made all the difference. We had people who were starting an ice cream business, a dog kennel, a memoir-writing service—Cliff didn't give a shit what kind of business we wanted to start, as long as we could prove to him that it had a shot at making money. He taught us how to read a balance sheet and to create a cash-flow projection, and he repeated his mantra over and over: most small businesses fail because of lack of planning and a lack of sufficient capital. If you don't have a good plan and enough cash, you're doomed.

As Gillian and I crunched the numbers and refined our business plan, we started to get a more realistic picture of the road ahead. We realized that even if everything went extremely well, we could never hope to make a huge amount of money from the farm. There was no way to justify borrowing a lot of money or investing a large sum—the returns just wouldn't be good enough. So we knew from the start that profitability depended on keeping the business as lean as possible. Cliff looked at our numbers and told me that we needed money not just to buy equipment and seed but also to live for at least two or three years while the business lost money. We had nowhere near the amount of capital we needed. Even with Gillian's consulting income, we didn't have a realistic plan to float the business and ourselves while we got established. Gillian and I have a tendency to charge ahead in the face of warning signs, and I was tempted to fudge our numbers to make it look like we were ready to start farming right away, but Cliff was watching me like a hawk. We simply didn't have a viable plan, and no amount of positive thinking was going to change that.

I was devastated. Our beautiful farm dream had smashed on the rocks of financial reality. Then an opportunity arose: Gillian's firm had landed a contract to build an internet portal for the government of Trinidad and Tobago, and they asked her to move south for a year to run it. When she refused, they sweetened the deal by offering to hire me as well, to write all the web content. So, on very short notice, we put all the farm plans on hold, packed up the family and moved to Port of Spain.

The decision to delay the birth of our farm for a full year was tough, but it turned out to be crucial for the success of our business. It allowed us to save some money to invest in the farm, but more importantly, it set the tone for how we would approach farming. We had created a rigorous plan and compiled good data, and then we had done what the numbers told us we must. We were dispassionate, realistic and thorough. Our hearts had told us to forge ahead, but our heads had set us on the proper path.

The timing of our year away also turned out to be fortunate. While we were gone, *The Omnivore's Dilemma* was published, and suddenly everyone was talking about sustainable agriculture. The *how* of farming suddenly mattered to a lot of people. The winter we missed was also the winter that never came. Temperatures all over North America were far above normal, and newspapers carried pictures of people golfing in January in Toronto. The crazy weather finally made people sit up and recognize that global warming was no longer an abstract theory—it was happening. By the time we returned from Trinidad, the word "sustainable" seemed to be on everyone's lips. Leaving the city and starting an organic farm was already a thing.

WE ARRIVED HOME IN EARLY MARCH 2007 and found the farm pretty much exactly as we had left it a year earlier. We had banked some money while we were in Trinidad, but not as much as we had hoped. We were still short, so we went all in and sold our house in Toronto. Now we had no escape hatch—we were committed.

Later that month, we drove down to Vermont to buy some crucial machinery. Eliot Coleman had said that a couple could manage five acres of vegetables with hand tools and a walk-behind tractor. A walk-behind is very much like a Rototiller—it has two wheels and a small gas engine, but it also has detachable implements. You can put a Rototiller attachment on it or any number of other things—a sickle bar for cutting hay, a wood chipper, even a mini hay baler. Walk-behinds are popular in many places in Europe, and the best ones are made in Italy. We found a place in Vermont that carried BCS, which we were assured was the Ferrari of walk-behind tractors. We bought the biggest model, thirteen horsepower, with a Rototiller, a bush mower and a snow blower. All in, it cost us over ten thousand dollars, as much as a small secondhand tractor would have. I was very uncomfortable spending so much money on a machine I had never even heard of a few months earlier, but Eliot told us to do it, so we did.

We hauled the BCS home in the back of a 2002 Toyota Tundra pickup that we bought from Charlie's Auto Village in Pelham, New Hampshire. It had 120,000 miles on it, manual windows and a cassette player, and it cost almost exactly the same amount as the BCS. It was big and dirty and smelled funny inside. It was the first and only pickup I have ever owned. I loved it so much it almost made me cry.

Buying a bunch of expensive equipment made the whole farming enterprise feel real for the first time. It was both exciting and terrifying. But before we started, we had to think up a name for the farm. At first we wanted something that evoked the place and the landscape we lived in, but just about every conceivable variation of "Creemore Hills Farm" had already been taken; weekenders have a habit of naming their farms and erecting fancy wooden signs at the end of their lanes. In the end, we called our farm what we had always called it. Foster had referred to my parents' land, with the little cabin on it, as "the farm" from the time he learned to talk, so both he and Ella called our place "the new farm." The more we thought about it, the more that name made sense. The New Farm would be the name of our farm, and of our business.

IN EARLY APRIL, WE STARTED BUILDING. The first thing we needed was a small greenhouse for starting our plants in the early spring, so I built one myself. Our financial prime directive was to stay lean, so I built a greenhouse that was very lean indeed. I constructed a rough wooden frame around our little back porch and tacked clear plastic sheeting over it. I found an old aluminum door in the barn and mounted that at the entrance. Visitors had to walk through our homemade greenhouse antechamber to get into the house. On cold nights we could open the house door and let some heat out into the greenhouse, so the seedlings wouldn't freeze. I thought it was an elegantly lean solution, though the whole contraption had an appearance that could in no way be described as elegant.

Eliot Coleman also told us we needed something that is alternatively called a cold frame, a high tunnel or a hoop house. This is where terminology becomes important. A greenhouse is heated. A cold frame looks exactly like a greenhouse to the untrained eye—it can be covered in plastic or glass or anything else that lets in the light—but it has no heat. A high tunnel or hoop house is basically the same thing as a cold frame but is covered in plastic, not glass. A hoop house simply traps heat from the sun—ours can be eighty degrees warmer than the outside temperature on a sunny day in February—so once it's up, it costs nothing to operate. A hoop house can allow a farmer to start growing earlier in the spring and continue later in the fall.

Eliot Coleman told us that we could build an inexpensive hoop house using electrical conduit for the frame, so in early April, as soon as the ground began to thaw, I picked up a quantity from Hamilton Brothers. I had the guys in the welding shop cut up some rebar into two-foot lengths, which I hammered halfway into the ground and slipped the conduit overtop, creating semicircular ribs at four-foot intervals. I bought a roll of greenhouse plastic from the Co-op in Bradford, the centre of the vegetable-growing area of the Holland Marsh, about an hour's drive from the farm. Hamilton Brothers also had bungee cord on a roll that you could buy by the foot (what didn't they have!). Once the plastic was stretched over the ribs, Gillian and I lashed it down with lengths of bungee thrown over the top, just as Coleman told us to. We thus covered a section of the original vegetable garden about twenty feet wide, forty feet long and ten high, and spent less than three hundred dollars on materials.

The next day was cold and sunny and calm. We planted salad greens, spinach, broccoli, peas and carrots in our new hoop house. It was only a few degrees above freezing outside, but inside it was beautifully warm and humid. We worked happily in short sleeves, sweating profusely. The day after that, the wind came up, or it might be more accurate to say that the incessant wind that blew pretty much non-stop from September to May resumed after a one-day hiatus. It was howling when I came downstairs in the morning. I looked out the back window with alarm. The wind gusts were causing the hoop house to alternately inflate and deflate, puffing up the plastic and then sucking it back down over the ribs. It looked like some kind of giant beast hyperventilating in the backyard, ready to take flight, tied down only by flimsy bungee cords. As I watched, a huge gust flattened the whole structure. Every part of our beautiful creation was suddenly in contact with the ground. I stood there in my pyjamas, mouth open, struggling to comprehend what I was witnessing, consoled only by the fact that it didn't blow away completely.

I spent the next week reinforcing the hoop house. First I bought more conduit and made a central ridge that attached all the ribs together, which made the whole thing stand up better. It snowed that night and flattened everything again. I dug several postholes down the centre of the interior and put fence posts in to prop up the conduit frame. When the wind came up again, the ribs slipped off the posts, which then stuck up through the plastic. Now my flattened hoop house looked like a decomposing whale that had somehow beached in our backyard, with bones sticking out through tattered skin. I finally went back to Hamilton Brothers and bought full lengths of rebar, took the structure apart and threaded the rebar through all the conduit hoops, then bought a new sheet of plastic. Coleman's lightweight, inexpensive plastic hoop house was now made of solid steel, had cost close to what I would have paid for an actual hoop house, and had taken me two full weeks to construct. But it finally held. And remarkably, the seeds we had planted in it started to grow.

ONE EARLY APRIL WEEKEND, about the same time that I was wrestling with the hoop house, our friend Gord called us up and asked if we wanted to go to the fur and feather show in Mount Forest. Gord lives near Duntroon, in a place he calls the Art Farm. His father was an artist, one of the first to move to our area in the early 1970s. Gord followed in his footsteps, attended art school in Toronto, and now lives and paints in the house he grew up in. He also raises sheep and grows a lot of vegetables. Gord's wife, Teza, runs a television production company. They keep a card on their fridge that their daughter Lilah made for a Thanksgiving project at school when she was very young. "I am thankful for my dad, who grows all our food," it reads. "And for my mom, who makes all our money."

Gord told us that the fur and feather show would be a good place to pick up some laying hens, which was high on our list of things to do that spring. I had never heard of a fur and feather show, but they're a fixture of small-town Ontario. Vendors sell and trade pretty much any kind of small animal—chickens, rabbits, goats, even peacocks and guinea fowl. The emphasis is on the weird and exotic, but Gord assured us that there would be lots of regular old laying hens for sale.

We had sold our minivan before we went to Trinidad, so Gord and his two kids picked up the four of us in his van early on a Saturday morning. Mount Forest is only a forty-minute drive from our place, to the southwest, and we got there before eight. It had snowed the night before, almost a foot of heavy, wet April snow, and everything was grey and slushy. We stopped and had breakfast at a greasy spoon on the main drag. Gord wasn't quite sure where the show was taking place, so we asked the waitress as we were leaving. "Oh," she said, "it's over at at the fairgrounds, but folks usually get there pretty early. It might be over by now."

That couldn't be right, I thought. It was a Saturday, and it wasn't even nine yet.

The waitress was right. We pulled into the fairgrounds as a stream of pickups drove out through the slush. Some of the stalls were still set up, rows of saggy tents with mostly empty cages and boxes. The animals that were left were very exotic indeed— bantam roosters with crowns of gaudy feathers on their heads, enormous rabbits the size of small dogs. One guy had a cage of ferrets; Ella went into spasms of grief when we refused to buy one for her. The vendors and patrons looked as strange as the animals. The whole place was like a rural version of a *Star Trek* convention. There was only one guy left who was still selling laying hens.

"You want females or mixed?" he asked. He was wearing one of those plaid Elmer Fudd hats with the ear flaps. I remember he had very few teeth. I looked at Gord.

"Females," Gord said. I tried to affect an expression that said, "Of course, I knew that. Why would anyone want a male laying hen?"

"They're pretty picked over by now," the guy said. "But I'll see what I can do."

His hens looked to be about two or three weeks old, about six inches tall, with rusty red feathers. They were all identical. The hen vendor grabbed a chick out of his cage and, with a dramatic flourish, held it high in the air so he could peer at its underside. Then he brought it down very close to his face, so the chick's crotch was within an inch of his nose, and squinted. "Nope!" he grunted, and shoved it back into the cage. He grabbed another one and repeated the performance. This time he decided that the chick was female, so he put it into an old cardboard liquor box. This went on for several minutes. I wasn't sure how he was keeping track of the rejects that he threw back into the cage with all the others, but he seemed to know what he was doing. Chick sexing is an art as much as a science. We've had hens and chickens for years now, and I have never once caught a glimpse of their genitals, male or female, despite the fact that the roosters spend just about every waking hour of their lives fornicating. Back then, I didn't have a clue, and I wasn't about to question an obvious pro. After checking all the chicks at least once, he had five in the cardboard box. "That's all I got," he said. "The rest are roosters." He charged us five bucks apiece.

We were all happy as we drove home, the kids in the back of the van peeking into the box of little hens. As spring slowly arrived on the farm, our little hens grew, and it became more and more obvious, with every passing week, that most of them weren't hens at all. We ended up with four very regal and very noisy roosters strutting around the barnyard, and one unfortunate, harried hen trying to avoid their affections.

THE LAST OF THE SNOW MELTED, and by mid-April the frost was finally out of the ground. It was time to start planting outside the hoop house, in our new garden. It had been a year and a half since the sod was turned, and we had applied copious amounts of Rubin's composted manure. Eliot Coleman and every other author we had read had drilled the organic mantra into our heads: "Feed the soil, so it can feed your plants." Most people think the difference between organic and conventional agriculture comes down to whether or not the crops are sprayed, but that notion largely misses the point. In organic farming, it's all about the soil. Conventional growers see soil primarily as a growing medium, something to hold the roots in place while they feed their plants with synthetic fertilizers. Organic growers focus on feeding and nurturing the incredible biodiversity in the soil, so that the soil, and the complex web of life within it, can provide the plants with what they need.

The best way to feed the soil and provide fertility, Coleman and the others explained, is to grow cover crops. These are plants that are grown not to harvest but to turn back into the soil. Different cover crops provide different benefits: legumes like peas and clover fix atmospheric nitrogen with the help of symbiotic bacteria that live on their roots; fast-growing crops like buckwheat smother and suppress weeds; and big, woody plants like sunflowers produce lots of organic matter to enrich the soil. Alfalfa fixes nitrogen and also has a long taproot that breaks up compacted soil and brings nutrients from deep underground up to the surface. Cover crops can be planted as monocultures or in mixtures (appealingly called "cocktails") that bring diverse benefits to the garden. We had planted buckwheat in midsummer the previous year on a brief trip home from Trinidad, and Rubin had ploughed it under before planting winter rye in the fall. Rye is a fantastic cover crop—it scavenges nutrients that would otherwise leach out of the soil over the winter, its thick roots hold the soil in place and prevent erosion, and it is allopathic, meaning it releases a substance into the soil that prevents weed seeds from germinating. I tilled under the rye with the BCS as soon as the ground dried out a bit. Then, after a week or two to let the rye break down, we planted.

We started with cold-weather vegetables like peas, spinach, lettuce, carrots and radishes. Lettuce seeds will germinate in soil that is just a degree above freezing, and the seedlings can tolerate a fairly hard frost when small. Our rows of tiny green plants were covered by a spring snowfall several times, but they thawed out and kept on growing. At the same time, the ramshackle greenhouse on the back porch was in full swing. Most commercial growers in North America start their seedlings in plastic cell trays, like the ones that plants come in at the garden centre—essentially large popsicle trays. Eliot Coleman told us to use soil blocks instead, so we did. Soil blocks are compacted blocks of potting mix, made with a special metal press that comes from Holland or somewhere. The seed catalogue we ordered our press from carried large, extremely expensive units that made twelve or sixteen blocks at a time, and cheaper ones that made four. We bought the cheap one. We used Coleman's recipe for blocking mixture—three parts peat moss, two parts compost, one part soil, some blood meal for nitrogen, a little crushed lime (the rock, not the fruit) and a lot of water.

Coleman enumerated the advantages that soil blocks enjoy over cell trays with a long list, my memory of which has been obscured by the uncountable hours I spent making soil blocks. I hated the whole operation. The blocking mix took forever to make and was heavy and hard to combine. If I didn't add enough water, the blocks would fall apart. If I added too much, the mix would stick in the press. The peat moss was seriously hydrophobic, so just getting the mixture wet was a chore. I spent whole days in the little greenhouse filling up trays of fifty blocks with my little press, four blocks at a time. The wet mix caused the skin on the ends of my fingers to crack and split. It was the kind of injury that looks so minor that I was embarrassed to even talk about it—I couldn't imagine a real farmer complaining to his friends about having little cuts on his fingers. But it was excruciating. Everything I touched sent a jolt of pure agony though my fingers. I suffered in silence, furtively applying Gillian's moisturizer to my fingertips when no one was looking.

We built shelves around the perimeter of our greenhouse and filled them with trays of soil blocks, then filled up the little dimples on the tops of the blocks with seeds. We planted trays of broccoli, cauliflower, cabbage, eggplant, cucumber and peppers, both hot and sweet. Gillian had found a seed company in the United States called Baker Creek that carried only heirloom varieties. Their catalogue was gorgeous, page after page of exotic veggie porn. She ordered an alarming variety of tomato seeds and planted close to a dozen trays, far more than I thought we needed. We also started many vegetables that normal growers wouldn't start in the greenhouse, because (why else?) Eliot Coleman told us to—things like sweet corn, snap peas, beets and spring onions that are normally direct-seeded into the garden. The greenhouse was soon a riot of growth, but it was still too cold to transplant anything into the garden.

BY THE MIDDLE OF MAY, WE WERE WORN OUT and dispirited. We had spent a year and a half planning, preparing and saving money, and what did we have to show for it? A yard full of ridiculous structures that made our home look like a *Beverly Hillbillies* set. A coop full of male laying hens. Bloodstains from my hemorrhaging fingertips on pretty much everything I owned. I felt like a fraud, like I was playing at being a farmer with my tiny garden and glorified Rototiller. I walked around all day under a cloud of doubt, burdened by the feeling that I didn't have a clue what the hell I was doing, which was probably a function of the fact that I didn't have a clue what the hell I was doing.

Worst of all, nothing was growing. Most of the stuff we'd planted had germinated right away, but then it just sat there. It was often fifty or sixty degrees during the day, but at night it would drop down to freezing or below. The seedlings didn't die, but they didn't look happy. We had planted a few bulbs of garlic the previous fall that had put up new shoots as soon as the snow melted, but they were now a sickly yellow colour. The stuff in the hoop house was doing better, but by the middle of May, it was still small. We had been spending money frantically for months, but our stunted plants showed no sign that they planned on producing anything we could sell.

After days of walking around the garden, wringing my hands, Gillian suggested we talk to her parents. They had grown a huge vegetable garden for decades. They would know what to do. I called, and Dick answered the phone. I was in full panic. Did I need some kind of organic fertilizer to get things going? What were we doing wrong? Why wasn't anything growing? Dick laughed. "The problem," he explained, "is that it's cold." A good organic soil, he explained, has lots of nutrients tied up in the living creatures that inhabit it. Those creatures aren't very active when the soil is cold, so the nutrients in the soil aren't available to the plants. "The yellow garlic is probably a phosphorus deficiency. It will clear up when the soil warms up," he said. "Don't worry, everything will start growing eventually. You just have to be patient."

That wasn't what I wanted to hear. Dick spoke with the calm, resigned air of someone who has grown food for more than half a century and has given up fighting the elements. I wasn't there yet. I wanted to do something now, not sit back and wait for a bunch of bacteria and fungi and worms to warm up and start doing their thing. But I didn't know what else to do, so I waited.

THE CREEMORE FARMERS' MARKET OPENS every year on the third Saturday in May. We had decided to follow Eliot Coleman's advice, and the prevailing orthodoxy, and sell all our produce in the market. We planned to start a CSA in our second season, once we had ramped up production and knew what we were doing. Our friends Jamie and Mark invited us to join them at their communal market table, and we accepted gladly.

A week before the first market, it was clear that we would have nothing ready to sell. I called Jamie and gave him the bad news. "I'm really sorry," I said. "I've tried everything, but stuff just isn't growing. We won't be able to join you for the first market." I felt like I was letting down the team, that I had failed in my first test as a farmer. Jamie laughed. "We never sell in the first market," he said cheerfully. "It's too fucking cold."

A few days later, just as Gillian's father had predicted, a transformation occurred. Our garden, which had languished for the six weeks since we had first started planting it, suddenly started to grow. The change in the weather was subtle (the furious west wind that had blown all winter slowly abated, the days got a little warmer, the nights a little less frosty), but the change in the rate of growth of our garden was anything but. Myriad factors determine how fast a plant will grow: soil temperature and fertility, moisture levels, the duration and intensity of sunlight. Different vegetables, and different varieties of the same vegetable, can grow at wildly different rates and will react to external stimuli in different ways at different times of the year. Whatever factors were at play in our first garden, some kind of tipping point was reached in the third week in May. My anxiety decreased. We were in business.

Gillian and I spent the weekend transplanting seedlings from the greenhouse out into the garden. The soil blocks demonstrated their advantage as we plopped the new plants into the garden. No fumbling with plastic trays or ripping roots out of drain holes. The cubes of soil just went straight into the garden, and the seedlings kept right on growing, without missing a beat. Our new garden was filling up, and we were so exhausted by all the work that we didn't even think about the fact that we were missing the first market.

By the next weekend, we had salad ready for harvest. We had followed the crop-mix orthodoxy that year and planted just about everything we could think of—dozens of different vegetables and multiple varieties of each. We didn't have an exact number, but at the end of that first season we estimated that we had grown about 150 varieties in total. Everyone we talked to, including Jamie and Mark, had told us that cut salads sell particularly well in the market, so we planted all kinds of salad greens, both under cover and out in the garden. Those we had planted in the hoop house were the first things that were ready.

I set my alarm for 5 A.M. on the Saturday of the second market. We didn't have any way to refrigerate a lot of salad, so I figured we would have to cut it immediately before we went to market, rather than the night before. Gillian had been up half the night with the kids, so I left them all sleeping. It was a glorious morning—cool, clear and dead calm. The sun was just coming up as I made my way through the half-empty greenhouse and walked across the lawn to the hoop house. It had slowly lost heat through the night as the temperature outside dropped, and now it was just as cold inside the hoop house as outside. The inside of the plastic was covered in condensation, and water rained down on me every time I bumped a support post or a rib. I knelt over the rows of salad greens and started plucking leaves with my fingers, to see how things tasted. I smile when I think of all the weird and wonderful greens we grew back then: Magenta Spreen, with clusters of bright pink and green leaves; Golden Purslane, a succulent with juicy stems and leaves that have a bright, lemony zing; Red Leaf Amaranth, with bright red leaf veins, that grows as a weed all over the world—we had eaten it as callaloo in the Caribbean and bonongwe in Malawi. There were also multiple varieties of mustard greens, two different arugulas and six or seven lettuces. The colder the weather, the sweeter the greens, so after weeks of struggling to stay alive as the temperature hovered around the freezing point, these guys tasted amazing.

I cut the leaves off just above the ground, using a pair of ordinary kitchen shears. Some of the varieties grew on a hard stem and had to be cut one leaf at a time. We had also planted some salad mixes that came premixed from the seed company. These were difficult to cut because the varieties in the mix grew at different rates and were all different sizes. The whole operation was slow and tedious, but I didn't care. It was a beautiful morning and I was undertaking our very first commercial harvest. I didn't recognize it at the time, but that was really the moment the farm was born—the moment when we stopped being gardeners and started being farmers. We were harvesting food for sale.

I put all the cut leaves into a plastic bin. When I had harvested everything that was ready, I took the bin outside and dumped it into a big cooler full of cold water from our well that I had set up on our back steps. Then I pulled the wet greens out, a handful at a time, and spun them dry in our little kitchen salad spinner. It took a very long time. Gillian and I had figured out this process only the night before. Post-harvest washing, handling and refrigeration was a major weak spot in our farm plan, an oversight that would come back to haunt us.

I ended up with about fifteen pounds of washed salad in the end, packed into another large cooler. I loaded an old wooden table, the salad and a cash box into the pickup and drove the ten minutes into town. The cutting and washing had taken me almost three hours, so it was after eight when I rolled into the market, which was in the parking lot of the Station on the Green, the community centre in the middle of Creemore. Mark and Jamie were already there. Jamie had a few pounds of cut lettuce, which he had washed in a plastic bin, then dried in a contraption of his own design—a milk crate with ropes tied to the handles that he would spin in a huge circle over his head. We estimated the weight and mixed Jamie's greens into mine. Mark didn't have anything ready yet—he was just there for the socializing.

It didn't take long to set up. We had the old table, the big cooler and a roll of plastic bags that Mark had stolen from a grocery store the year before. We had no tent, no sign, no straw bales or kitschy farm decorations. Just three dudes in their mid-thirties and their salad.

As we got ready, the first customers started wandering through the market. One stopped and stood in front of our table. Then another. It took me a minute to realize what was happening: people were lining up. The Creemore market opens at 8:30 sharp—there is a strict rule against selling anything before that time. So we stood there, the three of us looking at the growing crowd standing on the other side of the table, until we judged it was close enough to 8:30 to begin. Then we started bagging salad. We charged five dollars for a bag that held roughly half a pound, a price that Jamie and Mark had established the year before. The Creemore market had been going for about six or seven years back then, but there were only two or three vegetable vendors, all of whom were hobby growers. We were the only organic option, so our table was mobbed.

We immediately fell into a pattern that would hold, more or less, for the rest of that season. Jamie and I ran frantically back and forth to the cooler, bagging salad and making change. Mark worked the crowd. He chatted up old ladies, flattered young mothers and generally acted exactly like you would expect a small-town high-school drama teacher to act. Gillian and the kids arrived soon after we started selling, Gillian joining us behind the table and the kids wandering off to explore the market on their own. It was a beautiful, joyful morning. I felt like a kid, I had so much fun.

We had a line three or four deep until we ran out of salad just after nine. The market rules also state that vendors shall not take down their stand until the market closes at 12:30. So we scrounged up some chairs and spent the rest of the morning sitting behind the table, chatting with market-goers and other vendors. I met dozens of new people that day—weekenders and multigenerational residents, senior citizens and parents with babies in their strollers. The kids ran around with their friends from town and bought lemonade from a vendor at the other end of the market. Gillian and I went home with less than two hundred dollars in our pocket, but I didn't care. I finally felt that I had found my place in the community, that all my friends and neighbours had seen me in the market and said, "Ah! So this is how you fit in!" I felt something I had rarely felt growing up in the suburbs, or in all the years I spent working overseas. I felt like I belonged.

CHAPTER 4

THE GROUNDHOG WARS

BY THE END OF MAY, we had planted out all our seedlings and filled up just about all of our available land. Gillian's tomato plants went into long rows in the old vegetable garden behind the house, parallel to the hillbilly hoop house. She had started far more than we needed or could manage, but she couldn't bring herself to throw out any of the seedlings, so we planted them all. The last things to leave the greenhouse for transplanting were the broccolis, cauliflowers, cabbages and kohlrabis—all members of the brassica family. We grew several varieties of each that year and planted them in a grid, one plant every foot, at the farthest end of the new garden, near the gate at the end of our lane.

Gillian and I had adopted the habit of walking through the garden together first thing in the morning to check on everything. We called it our crop inspection tour. The morning after we planted the brassica seedlings, I noticed something that didn't look quite right. I thought we had planted the broccoli closer to the end of the garden, but there now appeared to be fifteen or twenty feet of unplanted ground down there. We walked over and knelt down in the soil. There was a tiny nub of a stem sticking out of the ground. I looked closer and saw little footprints in the fresh earth. We crawled around and found about that twenty plants had been chewed off, flush with the soil. One whole end of our garden was gone. "Shit," Gillian said. "Groundhogs."

Marmota monax, an omnivorous rodent of the family Sciuridae. Widely distributed in North America, common in the northeast and central United States and Canada, known by many different names. Growing up in Vermont, Gillian had called them woodchucks, which is also a derogatory term for an uneducated rural dweller. *Caddyshack* aficionados know them as gophers, and in some areas they're called whistlepigs. In Ontario, they're groundhogs. Our farm was lousy with them.

Our land had not been intensively farmed for almost two decades, with the exception of Rubin's plot of Roundup Ready corn immediately after we bought the place. With abundant grass and not many humans around, the groundhogs had multiplied. There were burrows all over, but especially in the fencerows, on the edge of the ditch along the sideroad, and in the thick cedar hedge that bordered our new garden. We had watched them over the past two summers, standing erect on their hind feet on the edge of their burrows or scampering through the grass when we went out for a walk, but we hadn't thought much about the impact of having so many large rodents living in close proximity to our garden.

Gillian planted another tray of broccoli that afternoon to replace the plants that had been lost. The next morning, we found that another twenty or so plants had been eaten. This made me extremely angry. We had spent a great deal of time, money and energy on those plants—making soil blocks, planting and watering, transplanting into the garden. These groundhogs were messing with our livelihood. There was an almost perfect arc of destruction at the far end of the garden, so it didn't take an advanced understanding of geometry to figure out where the culprit was coming from. I pushed back the branches of the hedge and soon found a substantial burrow, with signs of recent excavation. This bugger would have to go.

I retired to the house to do some research but found a dearth of information on organic groundhog control on the internet. What I did find seemed geared more toward home gardeners and involved fencing, something we didn't have the time or money to do. So I did what I usually did when my ignorance became an impediment to progress on the farm. I went to Hamilton Brothers.

Mike was behind the counter. I had barely uttered the word "groundhog" when he blurted out, "Shoot it!" I was beginning to see that this was Mike's fallback suggestion for just about every animal problem, but I wasn't ready to go there yet. I had a decent arsenal at the farm, old guns that friends of my dad had given to him and that he had in turn given to me. I had a nice old pump-action .22 that Bud Scully had found in his basement, and a .30-06 deer rifle with a scope that Keith Jewitt had hunted with for years before he got too old. And I had a semi-automatic 12-gauge shotgun that my dad had given me on my twenty-first birthday, at a time when I lived in downtown Montreal and had no interest in hunting. The problem was, you have to pass a test and obtain a permit to buy ammunition in Canada, something I had never done. I had only a limited supply of ammunition at home. Additionally, I had never actually seen a groundhog in the garden, and I didn't know if I could shoot one if I did. I asked Mike if there was a less lethal option.

I could tell Mike was disappointed in me, but he took me over to the area where they kept the live traps. These are just wire-mesh boxes with a spring-loaded door and a trip bar inside. I had first encountered a live trap in action at the Hamilton Brothers feed mill just across the road. The mill is run by Ted, a tall, lean man with a short white beard and a slow, shuffling gait. I've never seen Ted dressed in anything but coveralls, and he's perpetually covered in a fine layer of grain dust, which gives him a slightly ghostly appearance when he emerges from the dark interior of the mill. Ted has been in a protracted battle with the local raccoons for years. His office door has a handwritten sign on it that says, "Keep this door closed. Coons come in." There's always a live trap on the loading dock, outside the main entrance to the mill, usually baited with an apple. One day I arrived mid-morning to pick up some chicken feed and saw a raccoon in the trap, blinking in the daylight. "What are you going to do with him?" I asked Ted as we loaded sacks of feed into my pickup. Ted looked at me like I was stupid. "Shoot him!" he said emphatically. I later found out that common practice around here is to kill whatever gets caught in a live trap, by shooting, gassing with car exhaust or throwing the whole trap in a pond. It seems that the "live" in "live trap" is a bit of a misnomer.

Live traps come in different sizes. Mike told me to buy the biggest model, which I did. "What should I bait it with?" I asked him. I couldn't think why a groundhog would prefer some wilted vegetables in the back of a wire trap over the succulent seedlings growing right there in our garden. "Marshmallows," Mike said, without hesitation. I looked at him, slightly in awe. This man, I thought, knows everything.

I took a little detour on the way home, stopping at the gas station in Singhampton to pick up some marshmallows. The prospect of using marshmallows to catch groundhogs got the kids very excited. We put the trap right on the doorstep of the offending hog's burrow, tucked under the cedar branches. Foster and Ella went back to check if we had caught anything every five minutes for the first half hour, then lost interest.

The next day I woke up early and went straight out to check the trap. To my great surprise and satisfaction, there was an enormous groundhog in the cage. It almost filled the trap. I could hardly wait for the kids to wake up. When they did, the whole family went out in their pyjamas to inspect our catch. Groundhogs can be aggressive when cornered, but this one just sat there, even when I picked up the cage and set it on the tailgate of the pickup. Foster and Ella spent a long time inspecting the captive, with their noses almost right up to the mesh. They studied its sharp claws, long eyelashes and thick, shiny fur. They declared it cute.

After breakfast, the kids and I drove to the maple bush at the back of the farm to let it go. I wanted to go farther, but I didn't think it would be cool to release a groundhog on someone else's property. The release site was close to a mile from the garden, which I hoped was far enough. I didn't know if groundhogs had any kind of homing ability.

The release procedure for a live trap is a little weird. You pin back the spring mechanism on the door, then roll the trap upside down, which allows the door to fall open and the animal to escape. Our groundhog didn't like being rolled over very much, but once it found its feet it was off like a shot. Groundhogs are kind of fat, with short little legs. Running looks like an effort for them. This one just kept running and running after we let it go, in a straight line away from us. Foster and Ella found this extremely funny, and they laughed hysterically until it was far out of sight.

THE KIDS COULDN'T WAIT TO RESET the trap when we got back to the house. The euphoria of our first success came quickly to an end the next morning, however, when we found the trap empty, the marshmallows gone and a fresh swath of destruction in the garden. It was immediately evident that there was more than one groundhog at work, and that a light-footed varmint could get the bait while avoiding the trap's trip bar. The morning after that we caught something again, but it wasn't a groundhog. The kids came running in, terrified, to tell us there was something mean in the trap. I ran out and found a huge, muscular feral cat, hissing and spitting menacingly. It was a beautiful, healthy animal, and completely wild. I was scared to even pick up the trap with that ball of fury inside, so I locked the kids in the house and released it right there. I don't know if I have ever seen an animal run so fast. It went off like it had been fired from a cannon.

Over the next two weeks we caught and released one more groundhog, but there was also a lot of bycatch. It turns out that marshmallows are the universal bait, loved by all animals, wild and domestic. We caught the neighbour's dog, which was so big it made the trap look like a cube of fur. I could barely get the door open to let it out, it was so packed in there. We also caught our lone laying hen (Gillian speculated that it might have gone in there to find respite from the four randy roosters). Checking the trap became the morning ritual for the kids. They would delight in running back to the house to tell us what kind of animal we had caught that day. Then things got ugly. They tore into the house one morning, breathless and wide-eyed. "We caught a skunk!" they exclaimed in unison.

Skunks are strange animals. They're nocturnal and almost blind, and when caught in a live trap, they pace back and forth incessantly. I can't remember whom I called for advice that day, but they told me to throw a blanket over the trap before picking it up, because skunks won't spray in a confined space. I dressed in my shittiest clothes and some old gloves, trying to cover all my skin, then grabbed an old blanket and walked out to the trap. I had never seen a skunk up close before. It was smaller than I expected, and it didn't seem to see me at all. It just kept pacing back and forth in the cage. I threw the blanket over the top of the trap, then gingerly picked it up and placed it in the back of the pickup.

The kids were desperate to watch me release the skunk, so they jumped into the back seat and we drove to our usual spot at the back of the farm. I made them stay in the truck with all the windows up while I placed the trap on the ground with the blanket still over it. I wanted to be as far away from the trap as possible when the skunk got out, so I carefully unlatched the door, then found a long branch to roll the trap over with. I took a deep breath, pushed, and ran. When I got to the other side of the truck, I stopped and looked back. The trap had rolled over all right, but it had gone a revolution too far, landing right-side-up again. This meant three things: the door was still shut, the blanket was now *under* the trap and the skunk was pissed. The kids were in hysterics, laughing and shouting encouragement to me from inside the truck. I didn't see any other option, so I grabbed my long stick again and began rolling the trap over. The skunk raised its tail, stamped its feet and started aiming its backside in my general direction. I pushed, the trap fell onto its top, the door fell open and the skunk fired. It didn't get me too bad, but at such close range the smell was overwhelming. The kids were literally falling over with laughter when I got back in the truck.

"Daddy, you stink!" they shrieked as I started up the engine. The skunk waddled away slowly, completely unconcerned.

OUR TRAPPING PROGRAM HAD NO discernible effect on the rate at which our garden was being eaten. The groundhogs consumed the entire first planting of brassicas, then promptly developed a taste for lettuce and other vegetables. The level of destruction was starting to threaten the viability of our whole operation. It was now clear that I would have to take more drastic measures. So I called the one person I could think of who might be able to help: Donald McDonald.

Don, as I mentioned earlier, was my dad's best friend and the doctor who had delivered me. He and his wife, Pat, had retired to a ten-acre retreat with an old orchard they called Applewood, just up the hill from Duntroon, about ten or fifteen minutes north of our farm. Don was an obstetrician and gynecologist of some renown—he had spent much of his career at teaching hospitals in the Middle East—but deep down, he was a redneck. Actually, not very deep down at all. He loved to hunt, fish, drink and play cards, and he took great joy in tending his apple trees in his retirement. I often look to Don as an example of someone who managed to maintain a deep intellectual life but could also appreciate the pure satisfaction of manual labour and being outdoors. He lived in two worlds—one cognitive and scientific, the other tactile and practical—and thrived in both.

Don was raised in the tiny town of Ripley, Ontario, near the eastern shore of Lake Huron. My brother and sister and I had grown up with the four McDonald kids, and for years Don had regaled us with stories of his small-town childhood. His delivery was always completely deadpan, and I never quite knew whether he was bullshitting or not. He told us once about a troop of travelling entertainers who would come to Ripley in the summer to play donkey baseball. The rules were the same as regular baseball, but each player had a donkey that they had to hold on to at all times during play. Don and his friends would pay their admission, then howl with laughter as the players tried to field grounders or steal bases while dragging along a reluctant donkey.

I also remembered a story Don once told us about groundhogs. There was an old man, he said, who lived by himself in a run-down house outside Ripley. He worked for the roads department and kept a small supply of dynamite that he pilfered from the blasting sites he worked on. The old man loved to eat groundhog. Don would wax eloquent about the flavour of groundhog, how living in a hole and eating only grass somehow produced delicious, tender meat. Don and his buddies would shoot a groundhog, skin it and take it to the old man's house to trade for dynamite. I think Don said it was one stick of dynamite for one groundhog. Don and his friends would then use the dynamite for fishing—they would light a stick, chuck it into the local pond, then collect the stunned fish that floated to the top after the explosion. I always suspected that he was pulling our legs with that story, but it stuck in my head. Even if it wasn't true, I thought, Don will have some ideas about my groundhog problem.

If Don was pleased that I was consulting him on my groundhog infestation, he didn't give any indication. When I called him and asked him what I should do, he was all business. "Just pour a cup of gasoline into the burrow, wait a minute, then light a match and throw it in," he said. "But stand back." I didn't like the sound of that very much. I wasn't sure a single cup of gas would do a whole lot, and if it did, I couldn't stand the idea of burning groundhogs alive in their holes. "Don't worry," Don said. "The burning gas just sucks all the air out of the hole, and they suffocate. It's the most humane way to do it."

Gillian and I talked it over after I got off the phone. Our conclusion was fairly straightforward: the trapping program wasn't working, and we couldn't farm if all our vegetables were eaten as soon as we planted them. We decided to kill the groundhogs. So I grabbed the gas can and went to work. I poured one cup of gas into the burrow closest to the garden, waited a minute or so, then threw a lit match into the hole. There was a bit of a whoosh, then a bit of fire, but nothing too dramatic. The flames fizzled out after less than a minute. It didn't look to me as if anything fatal could have occurred deep underground. I had done some earlier research and read that groundhog burrows always have at least two entrances, and often four or five. The distance between entrances can be twenty-five feet or more. So I scouted around and found three different holes that seemed to be part of the same burrow. I poured a cup of gas into each, waited, then threw a match into the biggest hole. I turned, getting ready to ignite the second hole, when a huge fireball erupted out of the burrow. This was followed almost immediately by a concussive *boom* and jets of flame shooting out of the other two holes. I felt like someone had punched me in the chest, and my ears were ringing. Flames flared violently from all three holes, and there was a sound of rushing air, like a blowtorch makes. Now I knew why Don had warned me to stand back.

I flared four or five burrows that afternoon, but then I started a pretty major grass fire out in the ditch by the sideroad. Our most imminent groundhog problem was centred on the cedar hedge, but it was clear that I couldn't use Don's gasoline method in there—I would burn the whole thing down, for sure. To be honest, I took some maniacal pleasure in blowing up the homes of the creatures that were destroying my garden, but when I thought about it afterwards, the exercise seemed cruel and a little creepy. I decided to try something else.

The next day I hooked up a garden hose, stuck one end into a big burrow in the hedge and turned on the water, figuring the resident groundhogs would flee the rising water and escape out a secondary entrance. No such luck. It took several hours, but eventually the water flowed out of every hole in the burrow; all occupants were presumed drowned. Then I filled the holes with dirt and rocks, but the next day they were dug out again. I tried again and again, but no matter what I threw in there, the groundhogs always dug their way out. I hoped that they would get sick of all that digging and move on, but they didn't seem to. All the while, they kept attacking our garden, chewing seedlings and rows of spinach and lettuce down to the soil. I became desperate.

Groundhogs are diurnal, meaning they are most active in the early morning and late evening. We occasionally saw one out on the lawn during the day, but we had never seen one in the garden. I thought there might be only one or two problem groundhogs doing most of the damage, like the rogue elephants I had heard about in Africa that get into farmland and terrorize the villagers. If I could shoot a couple of the troublemakers, I thought, perhaps our problem would go away.

I got up before dawn the next morning, carefully loaded the .22 and crept out the back door, through the greenhouse. I immediately started seeing groundhogs, but they were all well out in the field, far from the garden. Groundhogs are territorial and never stray far from the safety of their burrow, so I didn't think any of them posed a threat to our vegetables. I snuck around to the front of the house, then froze. There was a groundhog right in the centre of the garden, upright and alert. He had heard me. I took another step and he dashed into the hedge. Damn. I walked over and peered into the cedars. I could see the hole he had gone down. As I stared, a head popped out of the hole. The groundhog was sitting there, staring at me. I slowly raised my rifle, and *crack!* I shot him right between the eyes. Groundhogs, I came to learn, are very curious. They will often come back out of their burrow immediately after being startled, just to see what's going on.

I shot five or six groundhogs in and around the garden over the next several days, including one impressive running shot with the 12-gauge after I had used up all the ammunition for the .22. Finally the damage began to abate. Don McDonald had made me promise that I would dress a groundhog for him so he could eat it, but I couldn't bring myself to do it. Handling a dead groundhog brings into clear focus the fact that it's a big, fat rodent, and the thought of skinning and gutting one turned my stomach. "Shame," said Don. "Grass-fed. They're such good eating . . ."

ALL THE WHILE WE BATTLED the groundhogs, the farm fell into a weekly rhythm centred on the farmers' market. We would plant, weed, water and prune, then get up very early on Saturday and harvest. As the season progressed, our stand had more and more to offer. The radishes were ready the second week: regular red ones as well as purple, long French and a pure white variety called Ping Pong. Spinach and early kale followed soon after, then garlic scapes and spring onions. Our market following also grew. Word got out that there was good organic produce at our stand but that we always sold out early, so people started to show up half an hour before the market opened. On the fourth or fifth Saturday of the season it was raining so hard when I woke up that I considered staying in bed. But I didn't want to let Mark and Jamie down, so I loaded up and drove into town. When I pulled into the half-empty market, very late, there was an orderly line of six or seven people in front of the place where our stand was supposed to be. There was no sign of Jamie or Mark. Our customers were all under umbrellas, waiting patiently in the pouring rain. Rather than being grumpy, they were grateful that we had come out in such terrible weather. That kind of loyalty had a deep impact on both Gillian and me, and we tried our best to repay it. We opened every Saturday, without exception, for the next five summers.

Growing for market, even at a very small scale, also taught us many things that home gardening hadn't. It taught us that the garden is a relentless taskmaster that will accept nothing less than diligent hard work, and that delaying a job almost always makes it harder. Weeding, for example, is fast and efficient when done with a hoe, when the weeds are very small, but if we got behind and the weeds got bigger, it took a lot longer. If we got far enough behind, we would have to weed by hand, which takes exponentially longer than using a hoe. If we got really far behind (which we sometimes did), the weeds got big enough to produce seed, meaning our future weeding workload would increase in that part of the garden. Some varieties of weed seed can stay viable in the soil for twenty years, so failing to weed on time can have an impact on workload that literally lasts for decades. We didn't learn this lesson all at once in that first season. But we eventually came to understand that certain jobs have to be done at certain times, no matter how exhausted or overworked we happen to be.

We also learned that growing under cover makes an enormous difference. Our hoop house was ridiculous in many ways, but it worked. The vegetables we grew under plastic were ready weeks before the same varieties planted outside, and our yields were far higher. A very large portion of the produce we sold in May and June came out of the hoop house. I began to fantasize about covering the whole farm.

Perhaps the most important lesson we learned that first season was about something called succession planting. We had read about this in Coleman's book and other places, but we didn't really understand how crucial it was until we started selling. Succession planting is pretty simple: a grower must plant repeatedly, on a regular schedule, to ensure a continuous supply of vegetables at the market. The problem is, as mentioned earlier, vegetables grow at very different rates, and those rates vary depending on the time of year and a host of other factors. An heirloom arugula planted in early April may take more than six weeks to mature. A hybrid arugula planted in midsummer takes less than three. We had made the classic mistake of planting everything in one shot at the start of the season, then waiting. We would end up with enormous flushes of produce that we often couldn't sell.

We planted a modest area of garden in beets in late May, for example, then were awash in beets at the end of July. Our market customers loved them—we grew traditional red beets as well as gold, white, cylindrical and striped varieties. Supermarket beets taste like dirt. A fresh organic beet is sweet and creamy and delicious. But after two or three weeks, all the beets were gone, and by then it was too late to plant more for that season. "When will you have beets again?" our customers kept asking. "Next year," I would mumble, looking at my feet. About that time I read an interview with a farmer in New York who had sold at the Union Square Greenmarket in Manhattan for years. He said that if he had to give one piece of advice to new market gardeners, it would be this: never stop planting. That line would come back to us, in different forms, from different sources, over and over through the years: when in doubt, plant more. You can't sell it if you haven't planted it. If you plant it, they will come. It was advice that would eventually transform our business.

IN EARLY JUNE, WE HAD OUR FIRST VISIT from the organic inspector. Gillian and I knew from the beginning that our farm would be organic, but we had a long debate over whether or not to certify. There are a bunch of different organic standards— some put out by independent organizations, some by national or regional governments—but most of them are pretty similar. All the standards are incredibly detailed and prescriptive—they list the products and methods that are and are not allowed, and include minutiae such as the precise temperature a compost pile must reach before it can be applied to a field, and the minimum width of the buffers required along the edges of an organic farm to prevent contamination from conventional neighbours. Inspection and certification are carried out by independent organizations accredited by the body that created the standard. When we started, the USDA standard had been around in the United States for several years, but the Canada Organic standard was just being rolled out.

We found that many small growers chose not to certify, for a host of reasons, some reasonable and some less so. Some of the farmers we talked to said they didn't certify because they didn't need to—they had a close personal relationship with their customers, who were free to visit the farm, ask questions and learn about how their food was produced. That made sense to us. But others told us that certification was too expensive, and that the record-keeping and reporting requirements were onerous. The more research we did, the less water that argument seemed to hold. We found a certification agency that charged three hundred dollars for our initial inspection and certification, and even today, with a much bigger and more complex operation, we pay less than a thousand dollars a year for our certification. The record-keeping requirements—what we planted, where we got our seed, what amendments we used—seemed like the kinds of things a good farmer would want to keep track of, regardless of whether or not they were certified.

In the end, we were swayed by an argument we heard from Ken Laing. Ken and his wife, Martha, have farmed organically near the north shore of Lake Erie since the 1970s, on land that has been in Martha's family for six generations. "I think it makes sense for farmers who are new to organic to certify," Ken said at a conference we were attending, "because it's very hard to know if you're really meeting the organic standards without the rigour of inspection and certification."

So the previous winter we had started the process. We filled out a lengthy application form and collected the required affidavit from Rubin that listed all the substances he had applied to our fields in the previous five years (just Roundup, and just that once). Then the certification agency assigned us an inspector.

Tom Cassan showed up on a sunny Wednesday morning. Tom was very tall and very thin, with a shock of wild black hair; he was a dead ringer for Ric Ocasek, the lead singer of The Cars. He drove an extremely beat-up Ford Focus, from which he emerged in a cloud of cigarette smoke. "Sorry I'm late," he said as we shook hands. "I like to poke around on my own for a bit before I start my inspection." Tom had apparently been snooping in our back field for half an hour before he drove into the lane.

Tom proceeded to walk Gillian and me through every inch of our operation. He rummaged through our soil-block-making ingredients in the barn (the blood and bone meal must be heat-treated), inspected the posts that held up our ramshackle hoop house (no pressure-treated wood allowed), and checked the label on the soap we used to clean the cooler we washed salad in (the organic regulations prohibit a whole range of chemicals found in cleaning products). The process was so thorough and invasive, my knee-jerk distrust of authority began to kick in, which I think Tom could sense. "I can tell you two are committed to organic," he said after a while. "I'm not trying to catch you doing something wrong, I'm trying to help you comply with the regulations." At one point Tom was making notes in a little book he carried. "What are you doing for groundhog control?" he suddenly asked, without looking up. I stammered. Gillian began earnestly inspecting her shoes. I tried to picture a list of acceptable varmint control methods in the organic standard that included shooting, drowning and immolation, but it seemed unlikely.

"Smith and Wesson?" Tom asked, a smile creasing the corners of his mouth. I nodded sheepishly.

"Perfectly acceptable," he said, and carried on taking notes.

After several hours of inspection, we went in the house and sat down around the kitchen table. Tom laid out our application, then brought out a paper copy of the Canada Organic standard—a binder about six inches thick. "I like to show this to people when they first apply," he said as he hefted the binder onto the table. "Everything you could ever think of doing on your farm, somewhere in here it will tell you if you're allowed to do it or not."

We spent a long time going through our application and discussing our growing plans, and it slowly dawned on me that for years I had been thinking about organic in the wrong way. I had always assumed that the primary objective of organic farming was to protect consumers from exposure to chemicals. We bought organic vegetables when we lived in the city so our kids wouldn't be eating pesticides. That's a big part of organic, but it's far from the whole story. Talking to Tom that day, I realized that the goal of the organic regulations is much broader: to create a sustainable system of agriculture that is separate from and independent of the chemical-industrial farming that dominates the landscape.

This idea hit home when Tom asked us about our seed supply. The organic regulations require growers to use organically grown seeds whenever they are commercially available. Now, from the point of view of the consumer, there is no difference between a head of organic broccoli grown from an organic seed (meaning it came from a plant that was grown organically) and a head of organic broccoli grown from a conventional seed. But for the food system, there's a big difference. By insisting that organic growers source organic seed, the entire organic sector is strengthened, more land is converted from conventional to organic, fewer pesticides enter the environment and fewer farm workers are exposed to dangerous chemicals. The goal is to create an organic food system, not just organic farms. "If we're going to displace conventional agriculture," Tom explained, "we need strong, profitable organic operations, from seed to plate. Buying organic seed is simply putting your money where your mouth is." The consumer is just one of the beneficiaries of organic food. The land, the environment and the farmers themselves have the most to gain.

Gillian and I were both exhausted by the time Tom climbed back into his car, lit a smoke and banged off down the lane. It felt like we had just survived the Inquisition. There would be several months of follow-up and requests for more information before we finally received our "transition organic" designation—then it would be one more season before we were fully certified. Now when I hear someone question the integrity of the organic label, I suppress my desire to smack them and calmly describe the ordeal of inspection, documentation and verification we went through that first year, and every year since. We all know that when it comes to food, terms like "natural" and "free-range" don't really mean anything. But "organic" is different. That year, we learned that "organic" carries the force of law.

Over the years I've become more and more certain that our decision to certify was the right one. Gillian and I have met dozens of growers who call themselves organic but freely admit that they don't follow the organic regulations to the letter. "I don't buy organic seed because I'm not certified," I've heard more than one "organic" farmer say. It's seldom out of a desire to dupe anyone, but without the rigour of the inspection and certification process, it's easy to justify cutting corners.

BY THE MIDDLE OF JUNE, our tomato garden was seriously out of control. The half million or so tomato plants that we had started in the greenhouse were now taking over most of the original vegetable garden out back—six or seven long rows of plants that were quickly morphing, *Jumanji*-like, into a solid mass of green. Most home gardeners tie their tomato plants to a wooden or metal stake, but it was so windy at our farm that we worried that everything would blow down. After doing a little research, we decided to invent our own system. I went to Hamilton Brothers and bought some cedar fence posts and a roll of wire. Mike hadn't heard of anyone using fence wire to trellis tomatoes (which should have been a red flag), but he thought that we should probably use high-tensile wire, to be sure it didn't break. My plan was to dig a post hole at each end of every row of tomatoes, then simply string a wire between them about two feet off the ground. I would then pick up each sprawling plant and tie it to the wire. And voila! Our tomato plants would be neat and upright, with minimal effort.

If only it were so easy. Fencing is one of those fundamental farming skills that I probably would have learned if I had grown up on a farm, but I hadn't. Mike had given me very little advice when I bought my supplies because it probably never occurred to him that someone who was attempting to run a farm wouldn't know how to put up a fence. The high-tensile wire was a nightmare to work with. It came in a coil about two feet in diameter. As soon as I cut the little wires that held the coil together, it sprang out, Slinky-like, all over the garden. When I managed to get it under control, it resisted all efforts to uncoil it in an orderly manner. There seemed to be some sort of perverse law of fence-wire entropy at play here—the wire was happy neither as a neat coil nor as a straight line between fence posts, only as a chaotic mess somewhere in between. When I did manage to run a wire between two of the fence posts, I couldn't get it tight. The wire was heavy and sagged a lot, and if I pulled too hard the fence posts began to fall over. It took me a very long time, but I eventually got all the tomato plants more or less upright and tied to a wire. It looked pretty sad.

Over the next few weeks, as the tomatoes grew, I strung lengths of wire higher up the posts. As the plants grew bigger and heavier, they would drag the bottom run of wire further and further down, until it was eventually lying on the ground. I kept adding wire and the plants kept pulling the wires down, until eventually I gave up. The tomato garden reverted to its previous state of wildness, only now there was more than a thousand feet of high-tensile wire intertwined with the mass of vines. I tried not to think about what it would be like to clean that up in the fall.

Our tomatoes started to ripen in early August, and by the middle of the month we were inundated. Even though the tomato garden was a horrible mess, the sheer number of plants ensured a copious harvest. We had tomatoes of every conceivable description: Black Krim, Green Zebra, Brandywine, Tangelo. We grew a variety called Fuzzy Peach, which looked exactly as the name suggested: pinkish, with fuzzy skin. Most of the varieties were delicious, but Fuzzy Peach tasted terrible. We had eight or ten varieties of cherry tomato, all different colours, which we sold in mixed baskets.

The biggest flush came in time for the third market in August. We harvested all day Friday, putting together more than 150 quart baskets of cherry and mixed heirloom tomatoes, as well as many bins of larger tomatoes that we planned to sell by the pound. It was already hot and sunny when we rolled into the market the next day before eight, and we anticipated record sales. I was slightly alarmed when Mark and Jamie started unloading bins of tomatoes as well, but our customers were already lining up. No problem, I thought. Everyone will want tomatoes on a day like this. Then, just as I was about to serve our first customer, another truck pulled up to our stand, one that I didn't recognize. A large, bearded guy jumped out, a little older than us. "Oh, this is Matt," said Mark. "I told him he could join us at the stand today." Matt Flett is a chef and avid home gardener who lives just south of town in Mulmur. He had once run a bistro on the main drag in Creemore, and he taught in the culinary school at a local community college. We're great friends now, but I didn't know him from Adam back then. Matt had brought only one thing to the market that day: tomatoes. Lots and lots of tomatoes. He unloaded tray after tray from the back of his truck, and my heart sank.

Our ridiculous oversupply of tomatoes was compounded by the fact that tomatoes are the one thing that almost every home gardener grows. "Oh, look at all your lovely tomatoes!" everyone would say. "But I can't buy any, I have so many of my own right now. Hasn't it been a great year for tomatoes?" Gillian and I went home that day with about three-quarters of the tomatoes we had harvested. After a morning in the hot sun, they were no longer saleable. That night, our roosters dined on about a thousand dollars' worth of beautiful multicoloured heirloom tomatoes.

Despite erratic supply, unpredictable weather and competition among members of our little market collective, our following grew, and Saturday mornings became the highlight of my week. Jamie and I would scramble around our stand when the market opened, selling our stuff, while Mark continued to chat up the customers. Mark would bring enormous zucchini from his garden, much bigger than anyone would want to buy, and then attempt to sell them to the most attractive female customers. His sales pitch was hilarious but he usually struck out; he tended to have more luck with the blue-hair set. Gillian would arrive with the kids shortly after opening, turn them loose and join us behind the table. We sold out every single week that season, usually by 10 A.M. Then we would hang out until the market closed, chatting with customers and friends. Lots of people came to the market with no intention of buying anything. They just enjoyed the socializing. We weren't making any money, but it was a hell of a lot of fun. One Saturday after the market, we all went back to Mark's house in Avening and had a potluck picnic made entirely from food grown by our three families. We sat around under the trees all afternoon, drinking beer and enjoying the delicious fruits of our collective labour, while the kids played in the Mad River. It was one of the most satisfying meals of my life.

AS THE DAYS GREW SHORTER and the nights cooler, our groundhog problem, which had festered all summer, suddenly roared back to life. Groundhogs breed in the early spring and give birth shortly after. Toward the end of summer they kick their offspring out of the burrow to fend for themselves. We were suddenly overrun by adolescent groundhogs looking for new homes. Some of them reoccupied the empty burrows close to the garden that we had blown up or flooded earlier in the season. Others started to dig new dens. One particularly audacious youngster spent several days trying to dig a hole right in the middle of our laneway, not twenty feet from our back door. I eventually shot it.

I resumed my early morning patrols and soon got the infestation under control again. I shot one holdout that had avoided me for days as it was sitting on top of a fence post at the front of our yard—groundhogs sometimes climb fences and trees to get a better view. But there was one individual that I saw repeatedly on our front lawn or in the garden that I couldn't seem to get. One evening in the late summer, we were eating dinner on the porch that we were in the process of building on the front of the house. It was hot and muggy and I had barbecued some venison that someone in the valley had given us. It was a little tough, but delicious. Ella was kneeling on her chair, chewing an enormous mouthful of meat while she looked out over the yard, when she suddenly froze. "Groundhog!" she whispered. The three of us looked over and, sure enough, my nemesis was out in the middle of the lawn, eating grass. My obsessive campaign against the groundhogs had rubbed off on the kids, and they had taken it upon themselves to scout for intruders. Ella turned to me, sweaty from the heat, her eyes wide, a wild expression on her face. She wasn't wearing any pants. "Shoot it, Daddy!" she almost screamed through her mouthful of venison. "Shoot it!" Meat juice was running down her chin. I looked over at Gillian.

"What have we become?" she asked in a defeated voice. I ran to get my gun.

THE GROUNDHOG WAS GONE by the time I got back to the dinner table with my rifle, but I saw it in the garden almost every morning after that. It was so skittish I never got a shot. By that time I had run out of both .22 and shotgun ammo and was down to the .30-06. A .30-06 is a large, extremely powerful and incredibly loud firearm. It's big enough to take down almost any large animal, save perhaps an elephant or rhino, so shooting a groundhog with it is literally overkill. I was always a little scared using it, but I had become obsessed with getting that last groundhog.

One day, I woke up at dawn for my usual patrol and looked out the window at the head of our bed on the third floor. There it was. The groundhog was out on the grass at the side of our house, about thirty feet from the back porch. I crept downstairs as quietly as I could. My parents had come up from Toronto for a few days and were sleeping in Ella's room on the second floor; Ella was bunking in with Foster. I went to the basement, unlocked my gun, loaded it and crept back upstairs. I gently opened the back door and stepped out onto the porch. We had taken the plastic off the back porch greenhouse, but the frame and all the shelves were still up and the whole thing was cluttered with empty trays, hoses and buckets of old potting mix. The groundhog was still there, grazing away, just a few feet from me. There was too much junk on the shelf in front of me to shoot over it, so I crouched down and laid the barrel of the gun on a crosspiece of the greenhouse frame. It was an awkward position, so I held the rifle with my trigger hand, using the other to steady myself. The groundhog was so close it filled my entire field of vision through the scope. I put the crosshairs between its eyes and slowly squeezed the trigger.

The deafening roar of the gun was accompanied by a sudden blinding pain in my forehead. I was thrown backward from my crouch, onto my ass, blood pouring down my face. My ears were ringing, and the pain in my forehead was like nothing I had ever experienced before. The recoil from my improperly supported rifle had driven the edge of the scope deep into my right eyebrow, opening a crescent-shaped gash about an inch long. I staggered into the house and ran into my mother, who had been awoken by what she thought was an explosion. I'm not sure what she thought when she saw her son walk in, still in his pyjamas and covered in blood, but she kept her cool and helped me clean up before the kids saw me.

At least I got the groundhog. The bullet had grazed the top of its head, removing the fur without breaking the skin, but the force from such a large round at such close range had been enough to kill it instantly. I had a lot of embarrassing conversations in the market that week, trying to explain how I got the huge, perfectly circular wound in my forehead. That was the last groundhog I ever shot. The following summer we bought a puppy from Gord at the Art Farm that the kids named Zip; he did a pretty decent job of keeping the groundhogs at bay. In later years our garden was big enough to absorb some losses from the odd groundhog anyway. We also realized that no matter what we do, wild animals will find a way to eat at least some of what we grow. It isn't worth the effort, the danger or the moral wear and tear that is required to kill them. In the end, we learned to coexist. We didn't have much choice.

CHAPTER 5

A WORLD OF PAIN

WE FINISHED OUR FIRST YEAR of farming with a grand total of $8,783 in gross sales, an achievement that inexplicably inspired us to try a second year. Gillian and I discovered much later that we had both come to the same conclusion after that first season: we would never make any money on the farm. But the only alternative we could think of was to sell the farm and run back to the city. That idea really stuck in my craw. We had started out with the intention of proving that we could make a living on a small organic farm. I thought about all the naysayers, all the people who had raised their eyebrows when we told them of our plans, our city friends and country neighbours who thought us crazy from the beginning. To quit after a single season would be to admit defeat. It would be humiliating. The thought of finding a real job and living in the city again was also horrifying. Gillian and I both loved our new community. We couldn't go back to the desk. We had to figure out a way to make it work.

One of the biggest reasons we decided to give farming another go, beyond simple pride, was a desire to put into practice all that we had learned during Season One. Our myriad missteps and mistakes had taught us a great deal, and it was painfully clear that just about every aspect of our operation could be improved dramatically. We had so many ideas about how to do things better, it was difficult to know where to start.

Perhaps the most glaring problem in Season One was a lack of infrastructure. Our goal of staying lean made sense from a financial perspective, but it severely hampered our productive capacity. We had no refrigeration, so we spent the whole season harvesting almost everything in the early morning, right before we went to market, which limited the volume of produce we could take and, therefore, our sales. This problem was compounded in the fall, when our garden was the most productive but the time between sunrise and market opening grew shorter each week; we simply didn't have enough hours of daylight before market to harvest everything. It was also ridiculous and embarrassing that I spent several hours every Saturday morning hunched over our tiny salad spinner on the back steps, drying salad a few handfuls at a time. So before our first season even ended, Gillian and I wrote off another chunk of the proceeds from the sale of our house in Toronto and came up with a construction plan.

First, we bought two secondhand hoop houses from a semi-defunct garden centre near Collingwood. The owner was a strange guy who repeatedly used the phrase "do me a favour" during his long and incoherent explanation of how to assemble the structures I had bought. "Do me a favour and bury all the edges of the plastic," and "Do me a favour and tighten all the purlins after you get the thing square." I wish he had done me a favour and actually included all the parts I needed, but at least the hoop houses were cheap: nine hundred dollars for the pair, cash, not including the plastic.

We decided to put up one of the hoop houses that fall. Mark and my friend Pete came up on a Saturday in late October to give me a hand. Pete and I had gone to summer camp together, but we had lost touch for many years. He is an artist and had recently rented Steve McDonald's place in Dunedin for a year while Steve, Jackie and the girls went off for a stint working in India. Pete is also an ultra-runner, and a pretty intense guy. He kept Mark and me entertained all day by dispensing wisdom on a wide range of topics, including gardening, personal health and the wonders of internet pornography. The hoop house was twenty feet wide and a hundred feet long, made of galvanized steel. We put up the frame in one day and left the plastic for the spring.

We also built a wash shed and bought a kit to make a proper greenhouse that would attach to it. Rubin came over with his excavator and removed the topsoil from the construction site, then I made a wood form for the floor and called in a concrete truck. Rubin showed me how to make a rudimentary oversized trowel out of an old broom handle and a length of two-by-six, and together we got the concrete floor reasonably smooth and level. We hired Gary Kramers to build the shed. Gary lives halfway between Dunedin and Creemore, along County Road 9. I first met Gary that fall when I looked out the window one evening and saw him stealing lumber out of my barn. When I ran out to confront him, he told me Rubin had given him permission to take it. (There was still some disagreement back then between Rubin and me over the ownership of the enormous amount of crap that he had left in our barn when he sold us the farm.)

Gillian and I designed the shed. It was about twelve feet wide and twenty feet long, with wide double doors at each end. The tall south wall was designed to support a lean-to greenhouse and had windows along the top that let in lots of natural light. Gary put the shed up during the winter and Gillian and I assembled the greenhouse in the spring; it was about ten feet wide and the same length as the shed, with clear polycarbonate panels and a cedar frame. We built enough shelves inside to hold over five thousand seedlings. The whole thing looked great.

Rubin's friend Russell Flack came over with his smaller excavator and dug trenches so we could run water lines to the new wash shed, the hoop house and the chicken coop in the barn. I bought some big secondhand stainless steel sinks, one from a farm auction and one from a restaurant supply store, and plumbed the shed myself. We also bought a used commercial double-door refrigerator. Now we were ready. We had a place to wash and process our harvest and a way to keep it cool. We had quadrupled our hoop house growing area and added a proper greenhouse that wasn't attached to the back of our house. All in, this new infrastructure cost close to five times our gross sales from the first season, but we tried not to think about that.

We next turned our attention to marketing. We never thought twice about returning to the farmers' market. We had developed such a loyal following and we had so much fun selling there, a second season was a no-brainer. But we needed other outlets.

If the small farm orthodoxy is clear about one thing, it's this: Thou shalt run a CSA. Community Supported Agriculture is all about sharing. Customers pay up front, before the season starts, so they share in the investment necessary to grow the food. During the season, customers get a box of fresh vegetables (and sometimes eggs, milk or meat) every week, sharing in whatever the farm produces. If a crop fails or some other calamity occurs, the customers get less, or nothing at all, so they share in the risks associated with farming. Some farms require CSA members to do a certain amount of work in the fields, sharing in the labour of growing their food. In theory, a CSA creates a lasting relationship between a farm and its customers, fostering a farm-centred community of people who care deeply about where their food comes from.

A very large proportion of small organic farms ran CSA programs when we started, so without thinking about it too much, we started making plans for a CSA of our own. We printed up beautiful rack cards with pictures of our produce and details of what we called our "farm share" program, and gave them out at the farmers' market. We planned to have pickup points in Collingwood, in Creemore and at the farm. In late October we got a cheque in the mail from a guy who lived in Toronto. We had sold our first share, to a complete stranger.

In November we sat down to figure out what seeds we would need and how much garden area we would have to devote to growing food for the sixty or so members we hoped to recruit. But when we began to hash out the details of what it would take to run a successful CSA, we started to get nervous.

The first problem was variety. We had grown an exceedingly large diversity of vegetables in Season One and had come to the conclusion that most of them were terrible. Some of them were too weird (Fuzzy Peach tomatoes) or too hard to harvest (salsify) or tasted bad (kohlrabi), and a lot of them people simply didn't want to buy (all of the above). We also learned that certain things grow better than others on our farm. Our sandy soil and cool weather are perfect for salad greens and root vegetables, but not so great for heat-loving crops like eggplant and melons. We had proved that we could grow lots of tomatoes, but we only had a two- or three-week window between our first harvest in August and our first frost in September. It didn't seem likely that our CSA customers would be happy receiving nothing but lettuce and potatoes every week all summer, but to grow *everything* successfully would be a daunting task. The thought of having to produce sixty boxes that contained a wide variety of beautiful produce every week for the entire season made my stomach tighten. I didn't think we could do it.

The other problem was personality. The goal of creating a community of CSA members sounded lovely, except that I would be forced to interact with a community of CSA members. Gillian and I are both reasonably social people: we have lots of friends, and we had grown to love our community of customers and fellow vendors in the farmers' market. But we also like our privacy. I wasn't at all sure that I wanted a lot of local food enthusiasts showing up at my farm on a regular basis. I pictured people with nose rings and long dresses who smelled of patchouli, wanting to engage in discussions of spirituality and gut health while I was trying to get work done on the farm. It's a terrible stereotype, but I'm trying to be honest. People often assume that because I'm an organic farmer, I'm into crystals and chia seeds, but I'm not. I have a low tolerance for fruitiness, and I like my space. Gillian is the same. A CSA seemed like a potential space-invading fruit-fest, so in our first rejection of a major tenet of the small farm orthodoxy, we decided to pull the plug on the whole idea. We refunded our sole shareholder's money and tried to think of another idea.

ODDS ARE YOU DON'T REMEMBER what the weather was like in the summer of 2008, but I do. The meteorological conditions that year are etched in my memory like a bad dream. It was cold and wet and miserable, at least on our farm. It snowed repeatedly in April and May. The wind blew so hard that it snapped many of our tomato plants in two when we planted them in the garden. We had three nights of frost in the first week of June. Our moods soon began to reflect the weather.

We had prepared a new, much larger garden the previous summer, to be ready for our ambitious expansion plans. Rubin had ploughed up about three acres to the west of the house, on the opposite side of the cedar hedge from our Season One garden. Eliot Coleman said that a couple could manage five acres using a walk-behind tractor and hand tools, with no outside labour, but we went for three, just to be safe. The new garden had been manured and cover-cropped and was ready to go, but by the end of April it was still covered in snow. I was so desperate to start farming that I finally put the Rototiller on the walk-behind and went out and tilled the snow. I thought it might make it melt faster.

When the farm finally did thaw out, we started planting like crazy, trying to make up for lost time. Our failure to succession plant had been a major error in Season One, so over the winter we had developed a fairly complex planting schedule to ensure that we would have a consistent supply of everything we were growing. Smartphones were a new thing back then—I bought a secondhand first-generation BlackBerry on eBay and we put our whole planting calendar into it. My phone would beep and tell me exactly when to plant.

We did almost everything by hand that year, using tools and techniques that we had read about in books or invented during Season One. Many of the basic methods we developed are the same ones we use on the farm today. The New Farm tagline is "Handmade Organic Food," and that's what we strive to produce. What did "handmade" actually mean back in 2008? Well, here it is in a nutshell.

First we prepared the ground for planting using the walk-behind with the Rototiller, the only step of the process that was mechanized. The Rototiller chewed up a swath of ground about thirty inches across, and it usually took a couple of minutes to make one pass across the two-hundred-foot width of the new garden, unless I was trying to take down a cover crop or some tall weeds, in which case it took a lot longer. The walk-behind was a big, heavy, violent machine that seemed to have a mind of its own. It weighed over 150 pounds, and with a thirteen-horsepower engine it seemed always on the verge of ripping my arms out of their sockets. It bucked and lurched and threatened to remove my toes if I wasn't vigilant, and it spewed exhaust into my face. I felt like I was continuously wrestling some kind of wild animal whenever I used it. I tilled the garden only a little bit at a time, but I calculated that to till the whole garden took about sixteen hours. I went over that garden four or five times that summer.

Next we planted. We had purchased an EarthWay push-seeder before Season One, a machine that was as simple and elegant as the walk-behind was brutal. The basic EarthWay design has been around for more than a century. It has two wheels, like a little bike, and a handle to push it. The larger front wheel drives a belt that turns a sprocket in the hopper behind it. The sprocket turns a seed plate that picks up seeds in the hopper and drops them down a chute. The seed plates are changeable for different-size seeds. The chute drops the seed behind a V-shaped wedge that pushes through the soil like the prow of a boat. The wedge is adjustable, so you can change the depth the seed is planted at. A loop of chain drags behind the wedge, filling in the furrow and covering the seed. The back wheel packs down the soil. An arm sticks out to the side of the seeder with an adjustable perpendicular rod that drags through the soil and marks the location of the next row. It took us a while to figure out the right combination of seed plate, planting depth and row spacing, but we eventually came up with a system that was fast, efficient and extremely consistent. We planted about thirty or forty rows of vegetables a week back then, which worked out to almost a mile and a half of row. We could plant that much with the EarthWay in a couple of hours.

We had learned in Season One that walking in a bed of vegetables, in between the rows, had a detrimental effect on growth, so we only walked in the spaces between beds. This meant that I had to hold on to the seeder with one hand, my arm awkwardly extended, to plant the middle rows of a bed. It also meant that the width of a bed, and the number of rows in it, was limited by how far we could reach with the seeder and how far we could reach when weeding. The layout of our farm was thus determined by simple human mechanics, not by the machinery we used. To grow by hand, the scale had to remain human. Salad greens were planted in rows about four inches apart, six or seven rows to a bed. We planted beets, carrots, radishes and the like in rows a foot apart, four rows to a bed.

After planting, we weeded. We had bought several collinear hoes on Eliot Coleman's advice. A collinear hoe looks like a letter T with a very long upright (the handle) and a very small top (the flat stainless steel blade). The blade is designed to be dragged along just under the surface of the soil, the sharpened edge cutting off weeds. Most weeds will die if cut below the root junction. The collinear hoe works extremely well on very small weeds that have just germinated, and it can be used with precision around close-planted crops like salad greens.

For larger row crops like beets or carrots, we used a wheel-hoe. The wheel-hoe works on the same principle as the collinear hoe, but the sharpened blade is mounted behind a single wheel and you push the contraption with a pair of handles that are offset, so you can walk beside a bed of carrots, for example, and cut down all the weeds between the rows. The wheel-hoe and collinear hoe dealt with weeds between the rows quickly and easily, but weeds that came up in the row with the vegetables could be removed only by hand. Our garden was brand new, and an enormous store of weed seeds had accumulated in the soil over the previous decades, what growers call a "seed bank." Weeding was a seemingly never-ending task. Gillian and I spent hours and hours on our hands and knees, pulling weeds. It was uncomfortable and tedious, and so fucking boring.

Some crops required special steps that involved even more labour. Cucumbers and tomatoes required careful trellising and pruning to avoid the overgrown chaos we'd experienced in Season One. Beets and carrots sometimes required thinning if we didn't get the seeder right and planted the seeds too densely. But the potatoes were the worst. We had no potato-growing equipment, except a furrower that dragged along behind the walk-behind and made a rudimentary trench in the soil. Into this trench we dropped seed potatoes and covered them up by hand, using a large hoe. Potatoes are grown in rows at least thirty inches apart, so there's lots of space for weeds to grow. When the potato plants were about a foot tall, we had to mound dirt up against them to prevent the developing tubers from being exposed to the sun and to kill the weeds, an operation that is usually called "hilling" but on our farm came to be known as "humping" (because it was funny). We again performed this task with a large hoe. We harvested by digging up the potatoes with a large garden fork. Every step of the process was heavy, back-breaking labour. It often took half a day to hump a single row.

Finally, after several rounds of weeding, we harvested. We cut salad greens by hand using large garden shears, and placed them in a plastic bin that held about ten pounds of cut greens. Salad greens wilt quickly after being cut, so as soon as a bin was full, we would walk from the garden to the wash shed and dump it into a sink full of cold water. Then we would drain and rewash the greens three times before spinning them dry. We had purchased a large bright orange hand-cranked salad spinner over the winter that held about a pound and a half at a time. Spinning was still hard work, but it was a big improvement over the little kitchen spinner. Then we packed the salad into plastic crates and piled them in the new fridge.

Pretty much all vegetables must be cooled immediately after harvest, or they spoil. The fastest way to remove "field heat" is submersion in cold water. We were lucky to have a deep well with abundant pure, cold water, though I didn't often feel lucky when my hands were in that water. We spent so much time with our hands in frigid wash water that Gillian began to permanently lose feeling in her fingers. We built a table with a top made of heavy steel mesh that we bought from Hamilton Brothers and used it for washing all our root vegetables. I spent so much time spraying down beets and carrots and potatoes that I was constantly wet and usually freezing cold, even at the height of summer.

Our new succession planting schedule meant that we were continuously restarting this cycle of labour—tilling, planting, weeding, humping, harvesting, tilling, replanting, and on and on. "Handmade Organic Food" didn't fully describe the kind of agriculture we developed. It probably would have been more accurate to say "food made through relentless, body-destroying physical effort," though that's not a very catchy tagline. I'm no stranger to hard physical labour. I paid my way through university on a tree-planting crew in northern Ontario, working ten-hour days, six days a week, hauling heavy bags of saplings over the most godawful landscape, swarmed by bugs and sleeping in a tent. But Season Two was the hardest thing I have ever done, by far. Gillian and I would fall into bed every night, exhausted, dispirited and aching all over. I would lie there, still vibrating from the walk-behind, and wake up with my hands locked in painful claws from gripping the handles of that wretched machine. I worked in the garden all night in my dreams, then woke up, stumbled out of bed, went straight out to the garden, and did it all over again. We had no time to see friends or go to the movies or do anything other than work. The farm became our whole world, and it wasn't a very happy place.

AS THE SEASON PROGRESSED and we planted more and more vegetables, our learning curve kept getting steeper. We had rejected many varieties that we tried in Season One, the ones we called "stupid vegetables," an accurate if uncreative name. Kohl-rabi was the poster child of the group—hard to grow, ugly, bad-tasting and no one wanted to buy it—but there were many others. Still, we were growing a very wide diversity on the farm. The kind of farming we were practising—many plantings of many different kinds of vegetables, staggered over the whole season—provided multiple opportunities to experiment and learn.

Our salad greens were probably the best example. We planted salad at least fifteen different times that season, and planted at least fifteen different varieties each time. We had at least three or four plantings of greens growing simultaneously at all times, all at different stages of growth, so it was easy to see what varieties grew best and the impact of different weather conditions. Salad greens grow fast, so when we tried different varieties or grow-ing techniques, we got feedback right away. We kept reasonably careful notes in a big diary and an ever-expanding catalogue of mental observations. I didn't realize how much expertise we were developing until visitors would drop by and I would show them around the farm. I'd start spewing information on the growth habits of the various vegetables in our garden and suddenly think, How the hell do I know all that?

I had always expected that as we got smarter and learned more on the farm, the work would get easier. How pathetically naive. Mother Nature repeatedly threw us curveballs, creating new obstacles that revealed the shortcomings and inadequacies of our growing system and forcing us to add more work. The cruel irony of our farm education was that the more we learned, the *harder* it got. I call this phenomenon the Flea Beetle Effect.

The crucifer flea beetle is a tiny black bug, a little bigger than the ball in a ballpoint pen. It tends to jump into the air when disturbed, like a flea: hence the name. The flea beetle is endemic to many parts of eastern North America and feeds on arugula, mustards, kale and all members of the brassica family, but it doesn't bother anything else. It prefers hot, dry weather but can appear anytime in the summer. There are lots and lots of flea beetles in most places in Ontario, especially on our farm. We first noticed the flea beetle in our garden during Season One. One day there would be none and the next they would be swarming all over the broccoli and cabbage and the mustard greens we grew for our salad. The bugs eat tiny holes in leaves and can stunt or even kill seedlings as they emerge from the soil. Conventional farmers deal with flea beetles with repeated applications of pesticides, but that avenue was closed to us. We tried an escalating series of measures, each one requiring more work than the last.

At first we tried to ignore them. Flea beetles spend part of their life cycle in the soil and appear in flushes throughout the summer. Sometimes there aren't any around. We thought that if we pretended they weren't a problem, maybe they would go away. That didn't work. The beetles tended to appear at the worst possible times, like when our kale was just germinating (they would eat the whole plant) or when the broccoli was just about ready to harvest (they would swarm all over the heads and make them inedible).

Then we thought about not growing anything that flea beetles like to eat, but that would have been stupid. Our customers and flea beetles had similar tastes.

Next we tried floating row cover, which is what Eliot Coleman told us to do. Row cover is a loosely woven fabric, usually white, that is draped over plants to protect them from insects; it's a common tool in organic agriculture. Row cover lets in sunlight, water and air, but the bugs can't get through (some grades also protect from frost). We bought it in big rolls, about four feet wide, and covered a bed of mustard greens early in Season Two, to see how it would work. It blew halfway to Dunedin the first night. We had weighed down the edges with rocks, as Coleman suggested, but that evidently wasn't enough on our windy farm. Gillian and I dragged the row cover back, untangled it and buried all the edges in soil with a hoe. Now it wasn't just potatoes that required humping. Every bed of kale and mustard greens and all our brassica plantings required covering, another expensive, time-consuming, laborious task. Long-season crops like broccoli and cabbage had to be uncovered at least once and weeded, then re-covered.

The row cover worked, mostly. We would often see flea beetles crawling all over the outside of the net, but when we carefully peeled back the cover at harvest time, the greens inside were perfect. Except when they weren't. About 20 percent of the time, the bugs would somehow find a way in. It's hard to see through the row cover, so we usually made the discovery when we took it off to harvest—mustard greens riddled with holes, completely unsaleable—and then had to scramble to find usable greens.

In later seasons we bought heavy-duty covers that completely eliminated flea beetles. These covers were very expensive, and they were too heavy to lay directly on the plants, so we had to add another step to the process—placing hundreds of flexible plastic hoops in the ground after planting to hold up the row cover. We finally mastered a technique that consistently produced extremely high-quality vegetables, but it required a system that was far more labour-intensive and expensive than we had ever thought possible when we started. The Flea Beetle Effect ensured that the better we got at growing food, the more complex and laborious our systems became and the harder we had to work at it.

THE MORE WE FILLED UP our big new garden, the more the neighbours took note of our activities. The long strips of white row cover seemed to especially pique their interest. People would often stop on the sideroad and peer out the open windows of their cars as we worked in the field. I would feel obliged to walk over and talk to them if I happened to be close to the road. "What are ya growing under all that plastic?" they would ask. I would explain that it was row cover, not plastic, and usually get caught up in a long discussion about our growing techniques. People were always very polite and interested in person, but there seemed to be a lot of private skepticism. Rubin had appointed himself official liaison between us and the traditional farming community, and he often gave reports on what "people" were saying about us. "Some people are saying, 'Who the hell do they think they are, this organic thing is a joke, they're just a bunch of dope-smoking hippies,'" he would tell us. "But most people are saying, 'They can do whatever they want, they're not bothering anyone.'" For some conventional farmers, the whole idea of organic agriculture was an affront, and calling us hippies was the gravest of insults. Mostly we seemed to be regarded as ridiculous but harmless. Everyone agreed that we would soon give up.

Rubin's wife, Lori, did her best to help us fit in. Lori is the thoughtful, well-spoken, attractive yin to Rubin's gregarious, loudmouthed, unshaven yang, but she didn't mince words when we unknowingly violated some local norm. She was particularly upset when she saw Foster and Ella playing in the barnyard in their brightly coloured rubber boots. "For god's sake, go to Hamilton Brothers and get those kids some black barn boots," she told Gillian in an exasperated tone. Lori thought that for a farmer (or their children) to wear rubber boots with fire trucks or tulips on them was akin to a banker showing up at the office in clown shoes. She also thought it necessary to point out the flaws in my field work. Farmers have always taken pride in their ability to plough a straight furrow, a skill that has become almost universally attainable in the era of GPS-guided tractors. Planting in a straight line makes all the subsequent tasks—weeding, hilling, harvesting—easier and more efficient. I attempted to eyeball my rows with the barely controllable walk-behind and the wandering EarthWay, and failed miserably. Our garden looked like it had been laid out by Jackson Pollock. Lori stopped on the side of the road one day, called me over and said simply, "You know, straight rows are important." I thought she sounded a little patronizing. I started setting up a string line for the first row every time I planted, so I could get it dead straight.

We were sometimes our own worst enemy when it came to fitting in with the community. Gillian has long been an enthusiastic practitioner of yoga, and she started attending a studio in Collingwood soon after we moved to the farm. In our second season, Gillian invited several instructors to hold yoga classes on our big front lawn, under the spreading maple trees. We would often have six or eight friends and neighbours out on the grass on a Wednesday afternoon, and I would sometimes reluctantly join in. I was usually the only guy. I knew that yoga provided many benefits, especially with the beating my body was taking on the farm, but I always felt a little fruity unrolling my mat with all the ladies. One afternoon I came up into Cobra and saw, to my horror, not one but two pickups full of McCormicks approaching on the sideroad. Rubin was driving the first truck and Lori the second, with all four McCormick boys fighting to get their heads out the window to see what was transpiring on our front lawn. Without thinking, I sprang up and tried to hide behind one of the big maples. Lori slowed to a crawl; both she and the yoga instructor looked at me with expressions of severe disapproval.

A few neighbours simply took pleasure in teasing us. Bill Watt and his wife, Martina, lived about a mile and a half down our sideroad, toward Dunedin. They had twins who were in Foster's class at school. Bill was one of the largest conventional farmers in the immediate area; he and his brother owned several hundred acres and rented several hundred more, growing huge monocultures of corn and soy. Bill thought our method of farming absolutely comical, and that the whole organic movement was a grand delusion, but he was always good-natured in his ribbing. One day he stopped on the sideroad in his huge pickup and urgently called me over. I trooped across the field and waded through the tall grass in the ditch to his open window. "Can't talk now!" he said joyfully, a huge grin on his face. "I've got to go spray!" Then he peeled off in a cloud of dust. "Tractor jockey!" I yelled at him as he sped away. Organic farmers have a limited arsenal of insults to hurl at conventional growers, and they're all pretty weak.

It was our friends in Dunedin, however, who saved us from despair as the work on the farm piled up. We hired Pete and Jackie to work a day or two a week, helping with weeding and harvesting. Jackie was a potato humper par excellence. Our friend Dan was always willing to help out with whatever needed doing, and Tara and Gillian became fast friends; they would often open a bottle of wine on a Friday evening and pick tomatoes together for hours in preparation for market.

THE SKEPTICISM AND OCCASIONAL HOSTILITY we felt from our immediate neighbours was more than offset by the love we received in the farmers' market. Jamie and Mark had retired from the market that summer, so in Season Two, Gillian and I were on our own. The pre-market lineups that had started in Season One continued, but because we were growing far more produce than in our first year, we didn't sell out so early, and the lines continued all morning. On long weekends we were thronged, with people four or five deep at our table from 8:30 until noon. Gillian and I would be frantic the whole time, bagging salad, making change and talking to our many regular customers as we ran around the stand. It was exhausting, but also tons of fun. Every week we had six days of drudgery and pain on the farm and one morning of unadulterated appreciation at the market. It was pretty much the only thing that kept us going.

But we also saw things in the market that reminded us how fucked up our relationship to food and farming sometimes is. Most of them were little things, comments or attitudes that were probably amplified by our fragile physical and emotional state, things that probably wouldn't have bothered me if I wasn't completely exhausted and overworked. The first was the demographics of our market customers. They were just about all rich. Quite a few of our friends bought from us—the young families who lived in Dunedin and Creemore—but most of our customers were weekenders or retirees who seemed to have a great deal of dough. People often showed up at our stand wearing swag from one of the private ski resorts or golf courses in the area, and it wasn't uncommon to serve a customer in riding boots and a hacking jacket, ready for a day of riding to the hounds. They were extremely nice people, but it grated on me that just about all of our food was going to the wealthy.

It was also clear that some of our customers didn't want to know too much about how their food was produced. "Is your stuff organic?" people would often ask. "Yes!" I would reply, then launch into a description of how we grew our carrots, excited to have the opportunity to make the farmer–eater connection I thought everyone wanted. But their eyes would glaze over and they would become visibly uncomfortable. Most people wanted to know that we were organic, but that was it. They wanted to check a box, feel good about their purchasing decision and move on. Partway through the season a new vendor started selling in the market. He put a huge banner on the side of his van that simply read "organic," then started unloading boxes of produce that were clearly labelled as conventional. Most of our regular customers commented that the new guy was obviously lying about the status of his vegetables, but others were oblivious and seemed to prefer to stay that way. They asked no questions; if they were being duped, they didn't want to know about it.

People also got upset when their farmer stereotypes were disrupted. My BlackBerry was a particular problem. People would often say things like "I've never heard of a farmer with a BlackBerry before!" They always sounded more than a little disappointed. It may be hard to believe now, but in 2008 a Black-Berry was a pretty serious status symbol. I felt the need to explain that I had bought it to manage our planting schedule, but that wasn't often seen as a reasonable justification. The simple truth was that many market customers thought not only that farmers are backward and uneducated but that they *should* be backward and uneducated. A farmer with a smartphone was obviously an imposter, or putting on airs. The irony of this attitude was that we were probably the most unsophisticated farmers in the county. I knew Mennonite farmers who had better phones than I did. Bill

Watt got up every morning and checked the commodities futures coming out of Chicago, ready to forward-sell his soybean crop up to two years in advance if the price was right, to buyers as far away as Japan. Half the land around Creemore was farmed by self-driving GPS-guided tractors that cost half a million dollars apiece. But many market customers clung to their stereotype of the hayseed farmer. They wanted to buy from someone simpler, dumber and poorer than they were.

But the most irritating market conflict was about price. We always displayed our prices on a chalkboard, but the vast majority of our customers never looked at it. They didn't care how much anything cost; they would pay whatever total we gave them without thinking about it. That was good, because I was painfully aware of the enormous amount of money we were losing, and of the terrifying amount of work that had gone into growing our market produce. People who complained that our stuff was overpriced were extremely rare, and I tended to meet their complaints with hostility.

The price conflict that stands out most in my mind involved Mrs. Black and some spring onions. Mrs. Black was very elderly and very proper. She and her husband had retired to Creemore, and she was a regular in the market. I would see her every Saturday, dressed in Mad River Golf Club gear from head to toe. I was never sure if she had just played, was about to play or simply wanted everyone to know she was a member of the swankiest golf club in the area. The first time Mrs. Black bought from us, she picked up a bunch of spring onions, looked at our chalkboard price list and asked in a disapproving tone, "Why are these two dollars? Foodland is selling spring onions for seventy-nine cents." Foodland is the grocery store in Creemore, immediately beside the market.

I felt my face getting red. Spring onions are a particularly fiddly crop. They germinate and grow slowly and put up thin, wispy leaves that do nothing to shade out competition, so they require a lot of weeding. We had priced ours at two dollars because we didn't think people would be willing to pay any more, but I was sure we were losing money at that price. Now Mrs. Black, a walking advertisement for a golf club that had a forty-thousand-dollar initiation fee, was complaining about being gouged. I took a deep breath and tried to explain.

"Our spring onions are much bigger and fresher than the ones they sell in Foodland. They aren't wilted, so you can use the whole thing, even the green part. Ours were grown organically, by hand. The ones in Foodland were drenched in chemical fertilizers and pesticides the whole time they were growing." Mrs. Black seemed willing to hear me out. "But most of all, they taste a lot better," I said. "I don't want you to pay me for these. I want you to buy a bunch of spring onions in Foodland and take them home, so you can compare. If you think ours are worth two dollars, come back and pay us next week."

The following Saturday, Mrs. Black walked up to our stand and handed me a two-dollar coin, without saying a word. She became a loyal customer and never again complained about the price of anything.

WHEN WE STARTED SEASON TWO, we spent almost every waking hour pondering a simple question: How would we ever grow enough food to make a living? All of our doubts and stress revolved around production; in order to make money, we had to grow more. So we worked in the garden until we broke ourselves. The relentless hours of gruelling labour, the constant financial stress and the omnipresent worry that everything we were doing was doomed to fail left me feeling shattered. The work and stress began to literally devour me; I'm six foot three and have always been skinny, but I lost weight rapidly, despite eating four or five meals a day. I bottomed out at 160 pounds by the middle of the season.

But as our skill increased and our efforts in the garden began to pay off, we were confronted with a new and even more difficult question: How would we ever sell everything we could grow?

In Season Two we came to the most important and transformational realization of our new career. Farming is made up of two separate but equally important tasks: growing and selling food. We were putting what we thought was a superhuman effort into growing beautiful food, but beautiful vegetables sitting in a farmer's field aren't worth squat. If our business was to survive, we needed to put just as much effort into selling that food. If you grow food, you're a gardener. You have to *grow and sell* food if you want to be a farmer. And we wanted to be farmers.

Our sales in the market more than doubled that year, but we had increased our overall production by a factor of five. All that extra produce had to be sold, or whatever effort we put forward in the garden would be meaningless.

CHAPTER 6

THE CULINARY WORLD

RESTAURANT SALES HAD BEEN in the back of our minds from the beginning, though we didn't really know why. Perhaps it was because the first small, diversified farm we ever visited—Runaway Creek Farm in Quebec—sold to restaurants. Perhaps it was because Gillian was a fantastic cook and a pretty serious foodie. Perhaps it was because Eliot Coleman gave restaurant sales his qualified blessing. Whatever the reason, we wanted to do it, but we didn't know where to start.

The small farm orthodoxy includes restaurant sales, but only under certain conditions. The image is one of a farmer showing up at the back door of the restaurant, handing over a few boxes of produce to the chef, and then heading back to the farm, cash in hand. Some farms might sell a progressive restaurant a few CSA shares. In the real world, almost all restaurants get their produce from a distributor. But the orthodoxy sees distributors as middlemen, and the orthodoxy doesn't like middlemen. Most organic growers see restaurant sales as a sideline, a small supplement to the real action of the CSA and the farmers' market.

We made our first restaurant sales during Season One by flogging our market leftovers to Chez Michel, the small French bistro on the main street in Creemore. The restaurant is run by Michel Masselin, who grew up on a farm in Brittany and speaks with a beautifully stereotypical French accent. Michel told me that his family had grown all their own food when he was young, an impressive feat considering he had ten siblings. They stored exactly one ton of potatoes in their root cellar each fall—enough to last the family all winter. We would show up at Michel's back door on Saturdays after the market closed and haggle over price, just as the small farm orthodoxy envisioned, but Michel has his own garden where he grows much of the fresh produce for his restaurant, and he's notoriously frugal, so he wasn't a reliable customer.

One Saturday in the early September of our first season, a very tall middle-aged man with close-cropped hair came up to our table in the market, toward the end of the morning rush. He was wearing a floppy felt hat with a large turkey feather sticking out of the band. He inspected our produce carefully, picking up vegetables, smelling them, turning them over in his hands. He seemed especially interested in our cucumbers. We had tried several varieties that year, some in the hoop house and others outside. The ones we grew outside were all terrible. The vines grew in a mess on the ground, the fruit was curly and misshapen and the plants were attacked mercilessly by bugs. The ones we grew inside did much better. Some varieties were very strange, like the Striped Armenian that grew to about three feet long and was covered in nasty spikes, but others were delicious. Our favourite was a Japanese variety called Tasty Jade that we still grow today. The tall customer picked up a cuke, studied it carefully and then, to my surprise, took a big bite out of one end. After a moment of contemplation he said, "I'll take them all" in a soft German accent. We loaded his bag with more than two dozen cucumbers. He paid his money and off he went. "That was Michael Stadtländer," Jamie whispered as he walked away.

Even I had heard of Michael Stadtländer, but I didn't know his whole story back then. Michael had grown up on a farm near Lübeck in Germany. He started cooking in a local restaurant while still in his teens and went on to work in kitchens across Europe, and as a baker on the German Navy's Destroyer No. 4 during his compulsory military service. In 1980, Michael met a young Canadian chef named Jamie Kennedy while working in Switzerland. They became friends and moved to Toronto to take over the kitchen at Scaramouche, which they transformed into one of the top restaurants in the country. Michael Stadtländer and Jamie Kennedy also established themselves as two of the most important and enduring names on the Canadian culinary scene, pioneers of modern Canadian cuisine.

Michael and Jamie went on to run restaurants of their own in Toronto, but Michael eventually grew tired of the gruelling schedule and the compromises he was forced to make in the high-pressure corporatized world he worked in. So in 1993, he and his wife, Nobuyo, a woman of tiny stature and enormous energy who came from Okinawa, Japan, left the city and bought a farm in Grey County, not far from the village of Singhampton. They called their place Eigensinn. It turned out that Michael and Nobuyo lived less than ten minutes north of our farm.

Michael bought from us fairly regularly for the rest of that first season. He almost always purchased a large quantity of one thing, and he never questioned the price. It didn't take a great business mind to figure out that selling in bulk at market price was a good way to make money. So in Season Two, we decided to go after the restaurant trade.

OUR TIMING, ONCE AGAIN, WAS LUCKY. The idea that chefs should cultivate a direct and symbiotic relationship with farmers and producers, an idea championed by people like Alice Waters and Michael Stadtländer, wasn't exactly new. But by 2008 it was starting to go mainstream. All across North America, organizations and individual chefs started figuring out ways to make the chef–farmer connection. Gillian started attending a lot of events in Toronto that winter. She went to something called the Green Carpet Series, where restaurateurs and farmers mingled while celebrity chefs worked with local products on a demonstration stage. She also attended a chef–farmer speed-dating event that was put on as part of the Royal Agricultural Winter Fair, where farmers sat at tables while chefs moved down the line, hurriedly describing to the producers what they were looking for. Gillian found it a bit comical. One side of the table was young, urban and cool. The other side was old, rural and decidedly uncool. One side had a lot of tattoos. The other side, not so much.

Gillian took the lead in our chef-courting activities, for a number of reasons. She was better able to function in an urban social environment than I was—she had nicer clothes and was usually more presentable. She "cleaned up better," as they say around here. Gillian also comes from a family with roots in the restaurant business. Her grandparents had run diners in small-town Illinois for many years, and her father helped run the New England Culinary Institute, one of the best-known cooking schools in the United States. Gillian had taken a bunch of culinary classes in high school and knew her way around a commercial kitchen. But mostly she just enjoyed schmoozing more than I did. Gillian is perhaps the greatest salesperson and promoter I have ever met. She has an uncanny ability to identify her desired outcome and then to make that outcome manifest, through sheer force of will. She claims that she has an unfair advantage as a brash American living among meek and timid Canadians, but I think it's more than that. Gillian can convince almost anyone of almost anything, and that winter she set out to convince chefs that they needed to buy our stuff. Just about everyone she met with agreed.

The first solid prospect to come out of those meetings was a guy named Ben Heaton. Ben has gone on to run a number of different restaurants, but back then he was the chef at the Globe Bistro, a local institution on the Danforth in Toronto. Gillian met Ben at one of the many events she attended, and he called several days later, asking for a meeting. "Would you come down and see me at my restaurant?" he asked. Yes, we would.

We made the two-hour drive to Toronto one winter morning, arriving at the Globe before lunch service started. One of the wait staff let us in and pulled a couple of chairs off a table for us. The place was dark and empty, and we felt like we were on the set of a mob movie as we waited for Ben to come out of the kitchen. He arrived a few minutes later, a ball of pure energy barely restrained. He spoke very fast.

"So, what will you have? Tell me what you plan on growing, and I'll tell you what I can use."

We went through our list. "We'll have several types of salad mix, radishes, spinach, multicoloured carrots and beets, lots of different potatoes, tomatoes . . ."

Ben checked things off verbally as we listed them. "Yes. Yes. No. Yes, we'll need lots of that. No, I hate that shit." His body language made it clear that we weren't talking fast enough. He jotted in a small notebook.

"What about delivery?" he asked impatiently.

"We'll be coming down once a week," Gillian said. That's what we had decided earlier.

"That won't work," he said matter-of-factly. "I have a very busy kitchen and limited cooler space. You'll have to come twice a week."

Gillian and I looked at each other, bewildered. "Okay," she said. "We'll come twice a week."

"Great," Ben said, and jumped up from the table. "Let me know when stuff is ready." Then he was gone.

Gillian and I sat there for a few seconds, trying to process what had just happened. Then we got up and let ourselves out of the empty restaurant. The whole meeting had lasted about four minutes. We paid for our parking, got in the car and drove the two hours home.

OUR FIRST MEETING WITH AN ACTUAL CHEF had brought into focus some of the logistical challenges of working with restaurants. Cooler space is always limited, chefs want their vegetables to be as fresh as possible, and no single restaurant goes through enough produce to keep a farm afloat. We would therefore need to make many small deliveries to many restaurants, several times a week. Both Gillian and I started to think to ourselves that spending at least two days a week driving—four hours' round trip from the farm to downtown Toronto, minimum, plus multiple deliveries—might not be feasible, especially when we already spent our Saturdays in the farmers' market. We were rapidly running out of days to grow food. But we kept our worries to ourselves, as usual.

We stumbled upon an alternative at an event that spring. In hindsight, it was probably the most important meeting we ever attended, though we didn't realize what a profound impact it would have on our business until much later. Organized by a non-profit called Local Food Plus, the event had the same format as a lot of other meetings, with farmers and producers at tables and chefs milling around making connections. It was billed as the biggest event of its kind ever in Toronto, so Gillian and I both went. A guy named Mike Schreiner was one of the founders of

LFP. Mike, who had recently been elected leader of the Green Party of Ontario, grew up on a family farm in Kansas and had run a number of food businesses in Toronto before going into politics. We had met him very briefly a year earlier through mutual friends. Mike saw us at the meeting and came over to say hi. We told him about our first season of farming and our plan to sell to restaurants. "How are you going to handle delivery?" he asked. "We're going to do it ourselves," we told him. Mike looked at us skeptically. He had been in the food and farming business long enough to develop some reservations about the small farm orthodoxy. "There are some people in the other room I think you should meet."

I have an image in my mind of that first meeting with Paul Sawtell and Grace Mandarano, founders and owners of 100km Foods. I see four people staring at each other across a table in the corner of a crowded room, mouths slightly agape, idiotic expressions on their faces. Four happy morons, sort of like a *Far Side* cartoon. My mental image probably isn't very accurate, but it captures our state of mind and the state of our two businesses back then. We were all clueless.

Paul and Grace had both worked for years as pharmaceutical sales reps. Through a series of revelations that I have never fully understood, they decided that plying doctors with free booze and junkets to the Caribbean so they would prescribe more meds wasn't a fulfilling way to live their life. So they quit and founded a farm-to-restaurant delivery business. When we met, they were actually more clueless than we were, if that's possible. We at least had one season of farming under our belts. 100km Foods existed only on paper at that point. They hadn't yet delivered a single carrot, and they didn't even own a truck.

We hit it off right away. Paul is upbeat and engaging, but with a cynical side that I really like. Grace comes from a large Italian family; she's funny and smart and very quick to laugh. Their business plan was simple. They would pick up directly from farms once a week and deliver to restaurants. All our products would be listed under our farm name, so chefs would know exactly where their produce was coming from. We would set our prices and Paul and Grace would simply add a markup. No haggling over price, no pressure to match the prices of other farmers. If two of Paul and Grace's farms were offering the same product, they would be listed and priced separately, and the chefs could choose. Most importantly, 100km Foods would handle all the billing and collection from the restaurants. We would know, when we put our product on the truck, that we would get paid. We knew that their plan for once-a-week delivery would be a problem for some chefs, but working with Paul and Grace seemed like a much better option than delivering ourselves. We shook hands and launched a business partnership that's still going strong a decade later.

THE FIRST THING WE OFFERED to 100km that spring was ramps. Also known as wild leeks, ramps are members of the allium family that grow wild all over northeastern North America and Appalachia. They are the first green thing to appear on the forest floor in the spring, and we had large, dense patches of them in the maple bush at the back of our farm. Ramps are delicious but incredibly pungent—all the old-timers around here have stories of eating raw ramps when they were kids, with the goal of getting kicked out of school for smelling so bad.

We called up Paul and Grace when the ramps were just about ready to harvest, sometime in the first week of May, and told them to put them on the list for the chefs. We had started eating ramps as soon as we bought the farm, and knew that digging and washing them is no easy task. They have thick, matted roots that must be delicately separated, and they cling to all the crap that makes up the forest soil—twigs, leaves, dirt, bugs. So we offered them at a price that we thought would deter most chefs—ten dollars a pound. With the 100km markup, chefs would be paying about thirteen bucks.

Back then, Paul and Grace were trying to work out their pickup and delivery schedule. The biggest challenge they faced was that chefs were accustomed to a produce distribution system that allowed them to order at the very last minute. Chefs would finish service late in the evening, look at what they had in their cooler, then call their produce guy, often at one in the morning, and put in an order, which would show up at the restaurant just a few hours later. Produce distributors often achieved this speed at the expense of quality and service. It was (and still is) not unusual for a delivery to show up with products that were substandard or missing altogether, or to substitute items without consulting the chef who had ordered them. Much like us, Paul and Grace started from a point of almost pure ignorance. They didn't know how the produce business worked, so when they decided that they would consistently deliver what their customers actually ordered, they didn't think they were doing anything radical. But they were.

100km wasn't big enough to carry inventory, so they had to wait until the chefs ordered to place the orders with the farms. This meant that we would get an order late Saturday night that we would need to harvest and wash in time to put on the truck Monday morning. That first Saturday night we waited excitedly for our order to arrive. When the email came in, I blanched. One hundred and twenty pounds of ramps. I had expected an order for ten or fifteen pounds. That was almost 15 percent of our entire first-season sales, in one order. How the hell were we going to do it?

By almost killing ourselves, that's how. We spent all day Sunday in the bush, on our hands and knees, digging ramps. It had snowed the night before, so the roots we were separating were a slurry of slush and mud and crap. It was incredibly slow and tedious. By working extremely diligently, Gillian and I could each pull about eight pounds of ramps an hour. We were dressed in insulated canvas coveralls and heavy jackets, but by 10 A.M. we were soaked to the skin and freezing cold. In the afternoon Gillian changed quickly and drove into Collingwood to find something to pack all the ramps in, while I continued digging. We had decided over the winter that we would ship all our produce in reusable plastic containers rather than the disposable, nonrecyclable waxed cardboard cartons that are the industry standard, but we hadn't gotten around to buying any. Gillian searched all over town and finally bought dozens of clear plastic sweater boxes from Bed Bath & Beyond. I honestly don't remember what we did with the kids that day. On many weekends they were left to their own devices while Gillian and I worked, both of us taking turns running into the house to whip up a meal or to check on them.

We finished digging just as it got dark and hauled all the dirty ramps back to the wash shed. Cleaning the ramps was almost as bad as digging them. We had to bring out pails of hot water from the house to thaw our frozen hands. At one point Gillian's fingers seized up and stopped moving altogether. It was like some kind of medieval torture, hunched over the sink, hands submerged in frigid water, washing mud off each individual ramp bulb. We fell into bed well after midnight, exhausted.

We still have the picture we took the next morning of our very first wholesale delivery. It shows Gillian smiling proudly by the open hatch of our Ford Focus station wagon, the cargo area completely full of white sweater boxes, five pounds of beautiful, clean ramps in each one. It was our biggest single sale ever, more than any day in the market the previous year. That first ramp order set the tone for Season Two and our new business of selling to restaurants—lots of demand and the potential to make money, but a degree of effort that didn't seem sustainable in the long term.

THE COLD, CRAPPY WEATHER THAT SPRING meant that it was a full month before anything else was ready to sell. We used the time to buy proper food-grade plastic bins to replace the sweater boxes we had shipped the ramps in. 100km Foods agreed to return our empties to us. We bought blue plastic bins with beige lids that became an important part of our brand. They distinguished our products from the other stuff (which all came in waxed cardboard boxes), they kept our product fresher and they were easy to stack and handle in restaurant coolers. Unfortunately, they were also great for holding all kinds of other stuff in a restaurant, and we often had trouble persuading the chefs to send them back.

The orders were modest at the beginning. We sold radishes and spinach by the bunch and salad mix by the five-pound bin. Paul and Grace had signed up about a dozen farms, both organic and conventional, that grew vegetables, fruit and specialty items like sprouts. We were the only organic vegetable farm they carried. We were also the farthest away from Toronto. Paul drove the truck back then, and he couldn't make it all the way to our place on his pickup run and still get back to the city before their warehouse closed (they were sharing warehouse space with a non-profit). So every week, I drove to meet Paul at Barrie Hill.

Barrie Hill Farms is a sprawling pick-your-own operation on the outskirts of the city of Barrie, a forty-five-minute drive east of our farm. It's run by Morris Gervais, a second-generation berry grower and one of the most enthusiastic human beings I have ever met. Morris oversees a frantic operation, with legions of summer students running the tills, retirees driving wagonloads of pickers back and forth to the fields, and about forty Mexican migrants working the fields. He had hooked up with 100km to sell his early-season asparagus and surplus berries, though I got the feeling that selling to Paul and Grace was a tiny slice of his operation.

Morris is one of those fast-talking entrepreneurs who can't help dispensing business advice to everyone he meets. The first time I went to rendezvous with the 100km truck, in June of Season Two, he gave me the grand tour of his operation and his opinion on a wide range of subjects. Morris grows conventionally. He was interested in our decision to go organic, not from a health or environmental standpoint, but from a business angle. We were charging three dollars a pound for radishes and eight or nine dollars a pound for salad mix, which he thought was both outrageous and excellent. "I can't believe the prices you're getting!" he said approvingly. He seemed to think "organic" was a code word for some sort of lucrative scam. Morris took justifiable pride in his operation and the food he produced, but he was in it to make money.

I began to look forward to my weekly trips to Barrie Hill in the same way I looked forward to Saturday mornings in the market. Both were a chance to socialize, but also the only moments when we made any money. It was fun to conduct business and experience some rare human interaction at the same time. Morris, Paul and I would load Paul's small refrigerated panel van, then stand around the tailgate and shoot the shit, talking business and farming and politics. Morris started calling this our "senior management meeting." Paul and I got to exchange a lot of ideas and information on our start-up businesses, and Morris happily assumed the role of mentor and dispenser of business wisdom, even though he was only a few years older than us. I learned a lot from those talks.

We also learned fairly quickly that chefs can be fickle. At some point that summer Paul and Grace came up with the brilliant idea of bringing chefs on the pickup run. I'm not sure the term "farm to table" had even been coined in 2008, but the chefs who bought from 100km that first season certainly cooked that way, and they were eager to see for themselves where their produce was coming from. One week, Grace told us that Amanda Ray, the sous-chef from Canoe, would be on the truck. Canoe was perhaps the most venerated restaurant in the city at that point. We badly wanted to get in there.

Grace said Amanda was looking for pattypan squash, something we had been growing since Season One. Pattypans are essentially miniature zucchini. They are usually star-shaped, but they come in all different shapes and colours. Gillian and I had been struggling to pick our pattypans when they were still small—Grace had told us that the chefs wanted them as small as possible—but pattypans grow extremely rapidly, and picking them at a very small size is a real pain in the ass. The plants are a jungle of huge leaves and spiny stems that cause a painful rash, and you need dozens of pattypans to fill even a small basket.

Gillian and I had gone out early that morning and selected only the smallest, most beautiful pattypans for the sample we were preparing for Amanda. They were stunning—tiny yellow and green orbs, perfectly formed and ridiculously fresh. Paul arrived at Barrie Hill a few minutes after I did. He introduced Amanda and I took her straight to the back of my truck to show her the pattypans. I carefully opened the bin, and she gasped. "Wow!" she said. "These are amazing!" I felt a flush of pride. "I've never seen such big pattypans before!"

WE DID EVERYTHING WE COULD that season to drum up restaurant business. Gillian cold-called chefs, we sent free samples out with Paul and Grace, and we attended every local food event going. We began to see many of the same chefs over and over again and began to identify a small group of what we called "true believers." Lots of chefs paid lip service to the importance of provenance, to the idea that where ingredients came from and how they were produced mattered. Lots of menus said things like "We source local ingredients whenever possible" or "We use organic products when available." In most cases, "whenever possible" meant "when it's cheap and easy, which isn't very often." The number of chefs in Toronto who actually walked the walk and went out of their way to develop a relationship with their producers was pretty small. It was this small group of true believers that kept our business afloat in the beginning.

The place where we ended up seeing the truest of the true believers most often wasn't a fancy reception or staged event in Toronto. It was just down the road, at Eigensinn Farm. Michael and Nobuyo ran a restaurant in their old brick farmhouse. In their living room they served a maximum of fifteen guests a multi-course dinner that featured meat and vegetables they raised on their farm and products sourced from a network of local growers. It was BYOB and extremely informal, but in 2002 Eigensinn had placed ninth in the influential San Pellegrino ranking of the top fifty restaurants in the world, one of only two Canadian establishments ever to make the list, and the only one ever to make the top ten. In Season Two we started delivering vegetables directly to Eigensinn. Unlike many other chefs, Michael and Nobuyo never called looking for something in particular. They would always ask what we had available and then choose what was freshest and most in season. We would drive it over, usually within a couple of hours.

We had adopted a strict "no delivery" policy once we hooked up with Paul and Grace, directing all prospective customers to 100km Foods. Eigensinn was the sole exception. I loved going over there to drop off vegetables, even if the size of the order didn't really justify the time and expense of doing it myself. The place was wonderfully chaotic. There were always chickens scratching around the dirt parking area between the house and the dilapidated barn, and pigs wallowing in the barnyard, just a few feet from where dinner guests would enter the house. The place smelled like a farm, not a restaurant. Beside the barn was an old school bus that Michael had converted into a mobile kitchen and dining room. He had driven it across the country several summers earlier, picking up local specialties along the way, then cooking dinner each night for whoever happened to be around in whatever town he happened to stop in.

I would carry in our vegetables through a laundry room that was always stuffed with crates of produce from the Eigensinn garden and Styrofoam coolers packed with fish and seafood that had been pulled from the Atlantic the day before and shipped overnight. The whole place had Michael's crazy, organic, free-form aesthetic. The kitchen had beautiful commercial equipment, but the island was built on top of what looked like a fieldstone wall, with little branches, animal skulls and old silver teapots embedded in the mortar between limestone and granite river cobbles. The place was built using a wildly creative mix of objects from the two places Michael loved the most: the kitchen and the outdoors. It was almost impossible to find a straight line anywhere on the farm.

We began to meet other chefs on our Eigensinn delivery runs. Michael and Nobuyo always had a half-dozen or so apprentices working for them, young cooks who wanted to learn how to raise animals, grow vegetables, make art and cook food, all at the same time. There were former Eigensinn apprentices in kitchens all over North America, and they often came back to the farm to visit the Stadtländers or to recharge for a night between the marathon stints of work that are typical in the industry. We met Kevin McKenna, who would take over at the Globe when Ben Heaton moved on, and Adam Colquhoun, the force behind Oyster Boy, and many others.

Michael and Nobuyo always had some sort of grand, often quixotic project on the go. That summer it was the Canadian Chefs' Congress. The idea was to hold a gathering of chefs from all across the country to promote the type of cooking that Michael believed in so passionately—organic and non–genetically modified, based on the best local ingredients and a partnership with those who produced them. First they planned to hold a fundraiser on the farm to underwrite the cost of the gathering later that fall. Gillian and I offered to help.

We arrived for the pre-congress fundraiser early on a sunny Sunday morning in July to help with the set-up. Gillian and I had both been to the Eigensinn farmhouse many times to make deliveries, but neither of us had ever walked the rest of the farm. Eigensinn was very different from the flat, fertile, open landscape of our farm. It was rocky and rolling and a lot more hard-scrabble; it would never have been productive cropland. We walked up a long, overgrown laneway from the parking area. The path was lined with wine bottles, tens of thousands of empties piled in a jumble along the road, like a fallen-down stone wall. The farm was criss-crossed with a network of paths through the scrubby fields, with strange art and idiosyncratic little buildings scattered seemingly at random. There was a huge statue of Bacchus made out of concrete and wine bottles, with an enormous magnum phallus. There was a half-completed guest cottage made of rammed earth and old tires, and a beautiful little log cabin in the woods where Michael said he had "schlumber parties." There was a statue of Mother Earth with a bread-oven belly that was accessed through an opening between her splayed thighs. The farm was a testament to unrestrained creativity.

The event was being held in the back hayfield. When we arrived, the chefs who had come up from the city to cook were busy setting up their stations, which were scattered across the field. Each chef had been given a steel tire rim and a grill that served as a barbecue, a table, a stack of firewood and nothing else. We helped several chefs get their fires going. "Is this normal for you, to cook like this?" I asked one of the chefs. He laughed. "No," he said. "Nothing's normal when you're working with Michael."

We met all kinds of amazing chefs for the first time that day. Mark Cutrara from Cowbell. Anthony Walsh, the executive chef of all the Oliver & Bonacini restaurants, who had built Canoe into one of the most iconic spots in the country. Jamie Kennedy used the arugula and heirloom tomatoes we had donated to make a mind-blowing BLT, with bread he had baked that morning and a thick slice of side pork grilled over the open fire. Everyone was incredibly nice, encouraging us to try what they were cooking before the paying guests arrived. All the chefs were donating their time, very evidently happy to be doing what they loved to do in such a beautiful setting. I felt like we were being welcomed into a second community, one built around a love of good food.

WHEN THE INVITATION TO THE FIRST Canadian Chefs' Congress arrived a few weeks later, Gillian and I got very excited. The meeting would take place over a weekend in September and was open only to industry—chefs, culinary instructors and producers. Registration was expensive, but with so many amazing chefs from all over the country meeting just down the road from our farm, we didn't think twice. The agenda also looked intriguing. Things were scheduled to get underway on Saturday, with the opening ceremonies at noon and the rest of the day taken up by three separate meals, back to back, culminating in a midnight barbecue. Lunch was scheduled to take four hours. We signed up right away.

My memories of the congress weekend are a happy blur. It was perfect late September weather, cool and sunny. Gillian and I drove over in the pickup on Saturday morning and set up our tent among the hundreds in Michael and Nobuyo's farthest back field. Chefs had flown in from every province and territory, and there was a happy, slightly chaotic air about the gathering, as if no one was sure exactly what was going to happen. The chefs marched into the cooking field in vague groups for the opening ceremonies, carrying their provincial flags tied to cedar branches that Michael's apprentices had pulled out of the woods. Gillian and I were the only farmers in attendance, so we just watched.

Lunch was cooked over open fires by the provincial and territorial delegations, thirteen stations spread around the hayfield. The Manitoba team made an incredibly delicious wild pike chowder. One guy from British Columbia was serving tiny oysters that his grandfather had originally brought from Japan, which were almost unbelievably rich and creamy, like little nuggets of foie gras—his family had been cultivating them in a single location on the coast for three generations. We ate fermented blubber and skin from a whale that the Nunavut team had harpooned in the high arctic. Michael and his apprentices marched in at the end of the meal with dessert, an enormous rustic cake with wild blueberry preserve that was about thirty feet long, served on a rough wooden plank. It took a dozen people to carry it. There was a seemingly unlimited supply of excellent wine from Niagara and the Okanagan Valley in British Columbia, and microbrews from all over the country.

Lunch merged seamlessly into dinner, which merged seamlessly into the midnight barbecue. At some point Michael made a speech and lit the bonfire, a pile of brush, scrap wood and slabs that was literally as big as a house. The heat was so intense that we all stood in a big circle about a hundred yards away from the inferno. The volunteer firefighters from Singhampton were on hand with a pumper truck, but they had hit the bar pretty hard and didn't look like they would be very effective if the fire got out of hand; luckily, it didn't. We all stumbled back to our tents well after midnight. We had been eating and drinking continuously for more than twelve hours.

That night, for the first time in months, Gillian and I went to bed without dreading the hours of work we would have to do the next day. We weren't stressed out about our precarious financial position or privately wondering if the whole farming idea was a colossal mistake. We zipped our sleeping bags together and cuddled in the cold September night, simply happy. Before we went to sleep, Gillian and I told each other that selling to restaurants was the best business decision we had ever made.

CHAPTER 7

CASTRATION 101

RAISING ANIMALS FORCES YOU to answer a whole host of questions that simply eating animals might not. Some are simple and practical: Where will they live? Some are huge and intractable: Is meat the product of murder? Our steep learning curve on the farm was mostly free of ethical pitfalls when it came to vegetables—our experiments in the garden didn't cause undue suffering for any lettuces. But as we had discovered with the Special Dual Purpose chickens, fucking up animal husbandry had serious repercussions for our animals, and for us.

The chicken euthanasia fiasco had made us gun-shy, but our roosters had done well and we had later increased our flock, adding fifteen or twenty laying hens in the middle of Season One. These we had bought as day-old chicks from a hatchery that evidently employed a more competent chick sexer than the guy at the fur and feather show, and they were all females. They produced exceptionally delicious eggs, with rich, dark orange yolks and a firm consistency that bore very little resemblance to the store-bought eggs I was used to. By the start of Season Two, we felt ready to take on more animals.

The small farm orthodoxy is big on livestock, which makes sense: meat, eggs and milk are high-value products that can bring revenue to the farm; animals eat garden scraps and surplus vegetables, turning waste into food; and composted animal manure is a high-powered, organic fertilizer. Livestock can also be integrated into long, highly sustainable crop rotations that build soil health and productivity on the farm. We considered all these benefits and one other: meat tastes good.

We had decided to raise meat chickens on a commercial scale way back when we created our first business plan for the farm, mostly because it was the only way we could make our numbers work. The profit we expected to generate with chickens was in no way based on reality, but it made the farm operation look more viable on paper. When we started to put our plan into action, however, things quickly fell apart. Poultry, eggs and milk are supply-managed in Canada, meaning who can produce them, and in what quantity, is strictly regulated. If you want to produce a supply-managed commodity, you first have to purchase quota, which gives you the right to produce a certain amount of that commodity.

I have no problem with supply management in theory. The Canadian system creates a stable domestic market and reliable supply, and it has been mostly good for the farmers involved. But the system is managed by provincial marketing boards that are dominated by farmers who already own quota, and they have powerful incentives to maintain the status quo. Until very recently, the minimum amount of quota that a new farmer could buy was very large, and very, very expensive. The marketing boards have also been reluctant to promote organic products.

When Gillian called the Ontario chicken marketing board, she was told that we could raise no more than a hundred meat chickens a year without quota, but the minimum amount of quota we could buy would allow us to produce something like ten thousand chickens a year and cost over a hundred thousand dollars. The woman Gillian talked to on the phone was unsympathetic. "We just want to raise a couple hundred birds, to take advantage of the market for free-range and organic chicken," Gillian told her. "I think free-range is disgusting," the woman snorted. "I don't want to eat a chicken that has been out eating bugs and god knows what. Chickens should be raised indoors." Gillian was shocked. "What's your name and address?" the woman suddenly asked. Gillian knew that we were legally obliged to register with the chicken marketing board if we had even one bird on our farm, but she didn't like the idea of this woman (or someone like her) showing up to check on us. "I don't think I want to tell you that," Gillian said carefully. That really made the woman angry. "It's people like you—" she started yelling. Gillian quickly hung up the phone.

WITH MEAT CHICKENS A LEGAL impossibility, we decided to go into eggs and pork. We bought a hundred laying hens at the start of Season Two (the most we could own without quota) and threw in a few ducks for good measure. Then we started preparing for pigs. Pork is not supply-managed in Canada—anyone can raise as many pigs as they please and sell them to whomever they want. We decided to ease into things and start with three.

The first order of business was a pen. The fencing around our barnyard was completely decrepit, so we hired Rubin to pull it all down with his excavator, burned all the old posts and took the wire to the scrapyard. Then Rubin put us in touch with some Mennonite guys who did fencing. The farmland in eastern Grey County, around Maxwell and Badjeros, was being rapidly bought up by old-order Mennonite families who had moved north from the Waterloo area, looking for affordable land. The first families had arrived a little before we did, and by 2008 there was a sizeable Mennonite community just across the highway from our farm. I talked on the phone to a guy named John. He quoted me a very good price and agreed to show up the following week with his crew.

The fencing operation was a two-day blur of frantic activity. John and his two helpers were very large guys dressed in typical Mennonite garb—straw hats, suspenders, white shirts and blue pants—but the way they worked and the equipment they used defied a lot of my Mennonite stereotypes. Everything arrived on a flat-bed transport truck (driven by a non-Mennonite), including two self-propelled fencing machines that looked like mini battle tanks. One machine drove in the fence posts while the other uncoiled and stretched the wire. John worked the piledriver like an extension of his own body, pivoting and swivelling the machine in mid-stroke to adjust the angle of the posts. The guys ate lunch in the barn out of five-gallon pails that were stuffed with sandwiches made with freshly baked white bread, and whole pies, washed down with multiple cans of Coke.

At the end of day one, John walked up to me and said, "We're done for the day."

"Great," I said.

He stood looking at me. "We'll be going home now."

"That's fine," I replied. We stood looking at each other for a few seconds, and then it dawned on me: John was expecting me to give them a lift home.

Gillian was in the city, so I piled Foster, Ella and three very large men into the pickup and made a ninety-minute circuit of Grey County to drop everyone. I had to wake the kids at five the next morning and bring them with me to pick them all up again. I never figured out why driving a fencing machine was allowed but a car was off-limits.

We ended up with a very tidy barnyard. There were two separate pens, both about fifty by seventy-five feet, one around the chicken coop and one next to it, farther from the barn, for the pigs. We also put up sixteen parallel runs of fence, each about sixty feet long and four feet apart, to grow tomatoes on. This replaced the failed tomato trellising system from Season One and was suitably sturdy, I thought, for our high-wind environment. It was all finished by the first week of May, right around the time we kicked off the season with our first ramp harvest.

My father-in-law had warned me that pigs have a Houdini-like ability to escape from their pens, so I spent a lot of time that spring creating what I thought would be the Alcatraz of pig-pens. I got some design advice from Mike at Hamilton Brothers, then stretched three strands of electric fence wire all around the inside of the pen near the ground, attaching it to the fence posts with yellow plastic insulators. I also ran a water line from the barn and built a little shelter in one corner of the pen so the pigs could get out of the rain and sun. When it was all finished, I plugged in the electric fence regulator. I couldn't resist the temptation to touch the wire, just to make sure it was working. It was.

GILLIAN AND I WENT TO PICK UP our piglets from Gerald te Velde in early June. I can't remember how we first got in touch with Gerald. He runs an organic farm near Meaford, about forty-five minutes northwest of our farm, where he raises sheep, pigs, cows and chickens and has a big market garden. He and his wife, Sherry Lynn, were about halfway through the production of a very large family back then. They now have six kids, I think, all boys except the youngest, and they're all home-schooled. Gerald has taken self-sufficiency to a daunting level.

A normal person would have brought the pigs home in the back of their pickup, but our truck used a lot of gas, so we avoided driving it when we could. We packed the three little piglets into a plastic dog carrier that we put in the back of our station wagon. The ride home could best be described as smelly.

Gillian and I had set up a bowl of water and food in the little shelter in the pen before we left. When we got the piglets home, we carried the dog crate into the middle of the pen, turned on the electric fence and opened the crate door. The piglets were each about a foot long, and very skittish. They came out of the carrier cautiously, but they startled and ran if we made any sudden movements. We stood outside the pen watching them, the kids oohing and aahing at how cute they were. Ella is a born animal-lover and extremely protective of all the creatures on the farm. "They don't know where their food is, Daddy," she said with concern. "I think they're hungry."

I should have left well enough alone, but I seldom do. I climbed back into the pen, over Gillian's objections, and tried to herd the pigs toward the food in their shelter. They started to panic, jumping into the air and running in short, extremely quick bursts. One suddenly made a break for the fence and managed to wedge itself between the two bottom electric wires. Pigs are pretty much cylinders of solid muscle. They're incredibly strong and extremely fast over short distances. This piglet was moving so fast that it was halfway through the fence before it felt the electric shock. Instead of turning around, it let out a high-pitched squeal and powered forward. It took only a fraction of a second and it was through the fence and out into the field. A pig had escaped from our heavy-duty Mennonite-constructed electrified pigpen in less than five minutes.

Back then, Rubin was still using large areas of the farm that we weren't. The pig fled into a green waist-high thirty-acre field of winter wheat that Rubin had planted the previous fall (one of the few things Rubin thought he could grow without spraying with pesticide). Finding a piglet in a wheat field is considerably more difficult than finding a needle in a haystack. At least a needle stays in one place. We all waded out into the wheat, but we couldn't see anything, and I feared we would just drive the pig farther away. The kids were distraught, but Gillian reassured them. "The piglet will get hungry and come back to its pen," she told them. It was evening, so we all went inside for dinner.

For the next three days we matched wits with the stray piglet and didn't come out looking too good. We would often see it close to the fence, communing with its siblings on the other side or drinking water from the bucket we had left out, but as soon as we got close, it would tear back into the wheat and disappear.

Finally, on the third day, we were standing around the pen with my friend Pete, who had stopped by to say hi, and we saw the piglet about ten yards into the wheat. It must have been getting pretty hungry by then, because it didn't run off right away. I circled around and got behind it, trying to scare it toward the barnyard. It suddenly bolted right at Pete, who, acting on pure instinct, reached down and pinned the pig to the ground as it tried to dart through his legs at full speed. The pig started squealing bloody murder and Pete started yelling, but he held it until I could run over and grab it by the back legs. I quickly heaved it over the fence and into the pen. That was the last time a pig escaped from our pen. Once they understood what the electric fence was all about, they stayed well away from it.

WITH THREE PIGS AND A HUNDRED laying hens, we started going through a lot of feed. Our pigs were a cross between two old breeds—Tamworth and Berkshire—that don't grow as fast as modern hybrids and that like to eat lots of things other than grain. The laying hens were also excellent foragers, but all live-stock, including heritage breeds, need some grain. Hamilton Brothers didn't have anything organic, so we contacted Gerald Poechman, an organic farmer in Hanover who makes and sells his own feed. The fact that Gerald is a third-generation egg producer whose last name is pronounced "Peck-man" made me eager to meet him.

Gerald is skinny, with dark hair and a big moustache, and he likes to talk. He started farming conventionally, like his parents and grandparents, but he switched to organic long ago and talks about organic agriculture with the zeal of a convert. He had about five thousand laying hens back then, a tiny operation by conventional standards, but big for organic. Gerald became our chief source of information on animal husbandry. A trip to his farm to pick up feed usually ended up being an all-day affair—an hour of driving each way and several hours of shooting the shit. He would run back and forth between the feed grinder and the grain chute, turning dials to adjust the mix of corn, oats, wheat, soy, flax and whatever else he had in his big grain bins. He grew much of the grain himself and bought some from other farmers, all of it grown organically.

Gerald explained that each animal requires different levels of protein and trace minerals in their feed. Too little protein for the hens and they would stop laying. Too much and they would lay enormous eggs that could cause their insides to spill out their back end. He adjusted our feed based on how we raised our animals (the garden scraps we gave our pigs provided extra minerals) and their age (older hens lay bigger eggs and need less protein). He asked lots of questions about our operation and was always wildly enthusiastic about our prospects, even when we weren't. "Don't give up!" he would say as I was leaving, slapping me on the back. "What you're doing is the future of agriculture!" I wasn't so sure.

Our animals grew rapidly, and soon our barnyard was an active, noisy, chaotic place. It was also highly sexualized. Two of the three pigs were male, and they began trying to mount the female pig, and each other, soon after they arrived on the farm, long before they were sexually mature. This was awkward and disturbing to watch, especially since they were all siblings. The roosters were constantly busy in a vain attempt to service all hundred hens. But the ducks were the worst. We had bought a dozen ducklings of several different breeds—Rouen and Muscovy and some big white ones that I can't remember the name of. They all grew extraordinarily fast and were evenly divided between males and females. By about six weeks of age the ducks had matched up into pairs, and the males were constantly chasing around after their mates, violently copulating with them over and over again. The males would also try to jump other females, looking for a quick extramarital fling whenever they could, which would result in noisy brawls that were half bar fight, half duck orgy. "What are the ducks doing, Daddy?" Foster asked, the first time he saw this going on. I had no idea what to say, so I went for straight-up honesty. "They're having sex," I told him. He looked disgusted, but it happened literally all the time, so the kids soon learned to ignore it.

Laying hens start laying at about twenty weeks of age. Toward the end of July came the egg deluge. A happy, well-fed hen will lay one egg a day, just about every day. Once all the hens got going, we were getting an 80 percent lay rate, which meant that our hundred hens were producing about eighty eggs a day. Gerald told us this was very good, but we didn't have time to revel in our success. We had to figure out how to unload forty-five dozen eggs every week.

In Ontario, eggs must be graded if they are to be sold anywhere other than on the farm. There were once egg-grading stations in just about every small town in the province (including on the main drag in Creemore), but consolidation did away with all that long ago. We discovered that the closest egg-grading facility was in Newmarket, more than an hour from the farm, so we decided to sell our eggs illegally. We took them to the farmers' market on Saturdays, keeping them hidden in a bag under the table. Our regular customers absolutely loved buying contraband eggs—some of them claimed it made them taste better. We also started a door-to-door sales circuit around town. I would drop by the offices of the *Creemore Echo* and the brewery, but our best stop was the bookstore, where the knitting club met every Wednesday afternoon; all the members had standing orders. I would sell six or seven dozen every time I went in there, the ladies all pausing from their knitting to stuff cartons of eggs into their bags of wool.

THE PIGS GREW AND GREW AND GREW. They gorged on overgrown lettuce and gnarly carrots and Gerald's beautiful organic grain. They especially loved beets, which dyed their lips bright red and caused their urine to turn a shocking shade of pink. The pigs were also very affectionate and they would nuzzle and rub up against us when we went into their pen to feed them. By August, I estimated that they weighed over a hundred pounds each. McCormick Meats doesn't kill pigs, so we booked them into a place near Mount Forest to be killed and processed at the end of September.

As the day approached, I began to think about how we would get the pigs to the slaughterhouse, a forty-minute drive to the southwest. I decided to build a pen in the back of the pickup, using old shipping pallets. The pigs were by then extremely heavy and powerful, so I built it strong. Then I made a ramp out of a sheet of plywood and propped it against the tailgate. It was one of those jobs that I became more and more skeptical about as I worked. I parked the pickup in the pigpen, set up the ramp, put the pigs' feed in the back of the truck, and waited. I could tell the pigs were hungry, but they flat out refused to go up the ramp. Even after we starved them for three days, they wouldn't go in.

I called Rubin to ask his advice. "They have some gates down at the plant that they use for sheep," he told me. "You could use those." I drove down to McCormick Meats to talk to Rick, Rubin's brother. Rick ran the abattoir with his brother Duane and their parents, Neil and Jean, and he raised sheep. Rick is a smart guy. He has a degree in agriculture from the University of Guelph, and he approaches farming with a thoughtful professionalism. I felt much better when he offered to help.

We loaded up several gates—lengths of wooden fence that can be pinned together to make a temporary corral—and drove back up to the farm. Rick said we would set up the gates to funnel the pigs into the back of the truck. But as soon as he saw our pigs, he stopped.

"Those boars are intact," he said.

"What does that mean?" I asked.

"It means they still have their nuts," Rick answered.

"Is that a problem?" It was pretty clear it was.

Rick explained that testosterone can cause boar taint. There's no delicate way to explain boar taint: the meat from an intact boar smells like piss when it's cooked. Standard procedure is to castrate male pigs as soon as they're weaned from their mother, but I hadn't known that. Rick told me that the government inspector at the abattoir would do a fry test after our pigs were killed—they would literally fry up a bit of the meat to see if it smelled bad. If it did (and Rick was pretty sure it would), the carcass would be condemned and disposed of.

It was late September. We were already beaten down and exhausted and bleeding money. The thought of losing two of our three beautiful pigs after all the time and effort we had put into raising them made my heart sink.

"What can we do?" I asked Rick. I was so upset I no longer tried to pretend that I knew what I was doing.

"Easy," he said, flashing a smile. "Just cut off their nuts." Rick explained that if we castrated the boars right away we could still have them slaughtered that fall. A month of testicle-free living, he assured me, would be enough to get all the testosterone out of their systems, and they would almost certainly be free of boar taint. I called the abattoir and put off the booking for a month, and then I called the boys in Dunedin.

BY THE FALL OF SEASON TWO, Gillian and I were comfortably ensconced in the Dunedin community. It was a diverse, slightly eccentric and extremely tight group of people. There were folks who had lived in the valley for multiple generations and newcomers from the city or other small towns. There were professionals with graduate degrees and people who had barely squeaked through high school. There were doctors and writers and teachers and cooks and cleaners. One friend was the North American sheep-shearing champion. Most of us were self-employed, which was good, because the castration was happening on a weekday; the vet wasn't available to help on the weekend.

I sent out word to everyone in the valley on Tuesday afternoon that we planned to castrate two very large pigs the next morning, and that beer would be provided. A half-dozen guys showed up. There was Steve, whom I had known since birth, and Dan, whom

I had met as soon as we moved in. Pete showed up—the escapee piglet had been named in his honour when he caught it back in the spring—as well as Jim and Jon. Jim is a farrier by training— he drives around in a van towing a portable forge, shoeing horses. He's also a chain saw artist. Jon was an ironworker until he fell and ruined his ankle; he now runs an industrial maintenance company. Jon is big and tall, has a long ponytail and rides a Harley. He looks like a total badass, but he's a softy inside. Gillian regularly beats him at tequila-drinking competitions at parties in the valley.

It was a cold, grey October morning when the guys showed up in their pickups and beater cars, a few at a time. Jason Durish, the large-animal vet, was the last to arrive. Jason had grown up in downtown Toronto, gone to vet school and never returned to the city. His son was in Foster's class at school, and they played soccer together. Jason had recently split with his wife and was perhaps the most eligible bachelor in Creemore (he said he sometimes woke up in the middle of the night to the sound of drunken women banging on his front door).

Jason quickly took charge of the situation. He explained that pig castration is done without anaesthetic, by cutting open the scrotum and pulling the testicles out. That was his job. The rest of us were tasked with keeping the pigs immobile while Jason worked. "The important thing to remember with pigs," he explained, "is that they're front-wheel drive. All of their power is in the front end, and they aren't good at backing up. So you need to keep their front end off the ground, to keep them from going forward." My palms started sweating.

We all stood around the pigpen, trying to figure out what to do. At that point the boars weighed close to two hundred pounds, and they were incredibly strong. My experience of trying to get them into the pickup had taught me that it was nearly impossible to force a pig to do something that it didn't want to do. I couldn't imagine trying to keep one still while Jason pulled its nuts out.

It was Jon who finally came up with a plan. The pigs' little shelter was built into one corner of the pigpen, with the entrance right beside the perimeter fence of the pen. Jon suggested that we herd the pigs into their shelter, then block their escape by tying one end of a twelve-foot farm gate to the fence to make a triangle with the base at the shelter door. Then we could push one pig at a time out the door, wedging it headfirst into the narrow end of the triangle. We turned off the electric fence and decided to give it a try.

Herding the pigs into their shelter was easy. They slept in there, so it was familiar, and we threw in a bucket of food to sweeten the deal. Once we got all three pigs into the shelter (we couldn't separate the female from her brothers), I stood in the doorway to prevent them from coming back out again. The others tied up the gate. Then I got into the shelter with the pigs to push one out into the trap. The shelter is only about six feet square, and very low. I was on my knees, crammed in with three very large pigs, but they were used to me and stayed pretty calm. I got behind one of the boars and started pushing, but as soon as he saw the set-up outside, he balked, turned, and wedged himself back into the shelter. No matter what I did, I couldn't persuade either boar to venture outside. I planted my feet against the shelter wall and pushed with the force of my entire body, but those pigs wouldn't budge.

"We'll have to snout them," Jason said. He found a length of rope, tied a slip knot on one end and made a loop, then handed it in to me. "Get that in its mouth and over the top of its snout, behind the canines," he shouted. I didn't like the sound of that, but the first boar put up surprisingly little resistance. I got a good look at his teeth while I worked the rope into his mouth and was suddenly a little scared. I handed the free end of the rope out to the guys, and they started pulling.

I had heard pigs squeal before that day. They sometimes let out a short, sharp squeal if they accidentally touched the electric fence, or if one was bitten by another. But I had never heard anything like the sound that pig made as we dragged it, inch by inch, out of the shelter. It was a constant, high-pitched, deafening scream. A snout rope works primarily by pain, the rope biting into the pig's nose and jaw. I pushed and the other seven guys pulled, and together it took us several minutes of intense tug-of-war to move that pig a dozen feet. It screamed the whole time.

Once we got the front end of the pig securely wedged in the triangle formed by the gate and the fence, we worked quickly. Jason jumped in behind the pig with me. Four guys each grabbed a leg from outside the fence or gate, and one guy kept pulling on the snout rope. "Get him into the air!" Jason yelled. Everyone pulled out and up, and the pig levitated a few inches off the ground. Now he really couldn't move—Jon's plan was working. Jason quickly squirted alcohol on the pig's balls from a squirt bottle like the ketchup bottle you find in a diner. A pig's testicles are high up on its backside and tight to the body. They don't hang. Jason and I were on our knees in the mud, with the boar's gear at eye level. Jason made two quick, expert slashes with the scalpel, and the pig went ballistic. It shrieked and roared and made sounds that I still can't get out of my head. All of us were yelling too, holding on to the pig for dear life, lost in the mayhem of the moment. Jason handed me the scalpel and, in a lightning-fast motion, reached in with both hands and pulled out a testicle in each. The mass of tubes and connecting tissue stretched momentarily, then came away. The pig's shriek died to a low, gurgling rumble. "It's done!" Jason yelled. We all let go.

The pig stumbled awkwardly and backed into the shelter with his siblings. I had hoped to untie the gate and let him out into the pen, but he didn't stay put long enough to do that. Everyone looked stunned. We were sweating and covered in mud, and several guys were visibly shaking. I felt like I was going to puke. The worst part was, we had to do it again. I crawled back into the shelter with the snout rope and tried to fit it on the second boar. The pigs were now understandably agitated, and they were moving around nervously. At one point I was pinned between the two boars and realized that, if they turned on me, I would be badly mauled, or worse, before any of my friends could get in there to help. I felt as if I deserved it.

I finally managed to snout the second boar, and the whole scene played out a second time. Every second of that morning is burned into my memory, but the most vivid is the sound—that awful, searing scream. Just remembering it makes me want to cover my ears.

It was not yet 10 A.M. when we finished, and a Wednesday morning, but we all opened beers and sat on the back deck for a while. "That was fucking crazy," Jim said quietly. Everyone nodded in agreement. We were shocked to see both boars acting completely normally within five minutes of their ordeal, rooting around their pen and nuzzling each other. I checked on them frequently for the rest of the day, but they seemed fine.

I replayed the morning's events over and over again in my head that night. The more I thought about it, the more I felt I had done something unspeakably cruel. I tried to console myself with the fact that just about every male pig in the world goes through the same experience, that every time anyone sits down to eat a pork chop or a slice of bacon, there's a 50 percent chance that it came from a pig that had had its testicles ripped out. But that didn't make me feel any better. I started searching online and found small farmers in the United States and the United Kingdom who claimed that boar taint was an invention of the industrial food system, that they had been slaughtering intact boars for years and the meat tasted fine.

Gillian found me on the computer after she put the kids to bed. "You should stop reading that stuff, babe," she told me. She put an arm around me, and I broke down. I couldn't stop thinking about what the pigs must have felt, the indescribable pain they must have experienced. My groin was aching. I cried while Gillian held me. "I've done a terrible thing!" I said, over and over again.

THE PIGS SEEMED CONTENT DURING the final month of their life. If the boars held a grudge, they didn't show it. They trotted over eagerly whenever I came to feed them, and they were as excited as ever to get a scratch behind the ears. I tried to treat them with extra kindness, just to let them know I was sorry.

That month also brought us to the lowest point in our farming career. After the brief euphoria of the Canadian Chefs' Congress, we descended into a state that was very close to despair. Gillian and I were both completely exhausted. The physical toil on the farm and the mental toll of perpetual financial stress had finally undone us. The weather got colder and wetter, the market ended, we made our last restaurant sale through 100km, and we realized that the whole year had been a financial bust. We had made a total of about $43,000 in sales, but our expenses were more than double that. With the fencing, wash shed, greenhouse, hoop house, produce bins, seeds, gas, chicks, piglets, feed and all the other millions of things we had bought, we were deeper in the hole than ever, despite not having paid ourselves anything. We had even spent a small fortune to dig a large root cellar for storing our beets, carrots and potatoes for winter sales. We would have been far better off financially if we had sat on the couch and watched TV for two years instead of running the farm.

As our bodies and minds broke down, Gillian and I began to turn on each other. We spent all day, every day, together. The farm was our whole life, and the line between our business and our marriage completely disappeared. We would talk about work during dinner, on the weekend, lying in bed. Arguing about how many rows of carrots to plant is the worst kind of foreplay.

Our fundamental dispute came down to this: Gillian said that we needed to grow the business if we ever hoped to make money; I said that if we grew the business any more, we would kill ourselves with work. The problem was, we were both right. But we walled ourselves into our respective positions and stewed in resentment.

I don't remember what issue finally put us over the edge. I do remember that we were outside on the lawn, near the little greenhouse, when it all went down. It had something to do with Gillian wanting to do something more and me not wanting to do it. Whatever it was, I snapped. "No!" I remember yelling. "No, no, no! No more!" I shrieked at her, then ran into the house. Gillian ran in after me. "What the fuck is the matter with you?" she screamed at me. There was no logic or coherent argument that day. We just turned all our pent-up anger and hopelessness and exhaustion on each other, and screamed and yelled and swore. It was the worst, most visceral fight we had ever had. We didn't talk to each other for days.

WE WERE FINALLY FORCED TO COMMUNICATE again when it came time for the final task of the season. We had to take the pigs to the abattoir. This time we borrowed our friend Gord's stock trailer. It was much lower to the ground than the pickup and had a proper entry ramp, but we still couldn't persuade the pigs to go in voluntarily. We finally resorted to the snout rope again. Gillian pushed and I pulled, and we eventually got all three pigs up the ramp. The pigs had topped out at over 250 pounds each, so it took superhuman effort to drag them into the trailer. They screamed the whole time. The sound inside the steel trailer was overwhelming. By the time the last pig was in, Gillian and I were both soaked in sweat and crying like babies. We sat down in the wet grass outside the pigpen and told each other we were sorry.

We took the pigs to an abattoir that was run by Mennonites. It was small and clean and professional, but like all abattoirs, it was a horrible place. It stank of shit and blood and was filled with the sounds of terrified animals. We backed the trailer up to a loading dock. A worker got in and whacked each pig with a steel rod that had a series of pins on the end, tattooing the pig through the skin so the abattoir would know which carcass was which after the pigs were killed; this hurt a lot, judging by the sounds the pigs made when hit. Then the staff chased the pigs from the trailer and down a run into a holding pen. Our pigs had spent their whole life on grass and had never experienced a concrete floor before. They slipped and skidded, their panic rising with the sounds of other pigs in distress. When one pig tried to head back into the trailer, the worker hit it with an electric prod. The pig screamed, fell over, then scrambled up again and raced away to its doom. I felt a sickening remorse as I drove away.

The next day I called my parents in Toronto, just to say hi. My dad answered. "What's the matter?" he asked. "You sound a little down." I told him that the farm was losing money faster than ever, that I didn't see any way we could turn things around, and that Gillian and I had been fighting. "I'm starting to think this whole farm idea is stupid," I told him. "I feel like maybe it's time to cut our losses and give up." It was the first time I had said those words out loud.

My dad is a pathological entrepreneur. He dropped out of high school, left his native New Zealand and worked his way around the world before falling for my mom and settling down in Toronto. He has been self-employed for most of his life, and has always had to struggle to make his own way. He sees the world through the prism of money. I never got the impression that my dad thought farming was a good idea. That day, when I told him I was thinking of throwing in the towel, I expected him to agree with me. But he didn't.

"No," he said firmly. "You've invested too much and come too far to give up now." I guess experience had taught him that starting a business requires prolonged effort. "You have to give it one more year." It sounded like an order.

Gillian called her parents the next day and had exactly the same conversation. They both got on the line, as people of their generation like to do, and implored her to carry on.

So Gillian and I had a formal year-end meeting, the first of what became an annual tradition. We sent the kids off to school, then sat down across from each other at the kitchen table. "If I have to choose between our business and our marriage, I'll choose our marriage," she told me. I agreed. It took us most of the day, but we worked out a compromise between her desire for growth and my desire to stay alive. In the end we decided to give it another go. There would be a Season Three after all.

CHAPTER 8

LEARNING TO WORK

ELIOT COLEMAN SAYS that a couple can make a living on two or three acres, using nothing more than a walk-behind Rototiller, with no outside labour. Gillian and I followed that prescription for two seasons and came to an irrevocable conclusion: Eliot Coleman is a liar.

In late 2008 we realized that if we ever wanted to make money on the farm, we would need more of everything—land, labour, machinery, the works. So we moved aggressively. We paid Rubin to come over with his big tractor and work up a new five-acre field, just to the east of the house. He then brought in about twenty dump trucks full of composted manure that he had been stock-piling at one of his farms. Rubin spread it all with the biggest manure spreader I've ever seen. Manure spreaders are universally known around here as "shit spreaders," and man, did that thing spread shit. It had huge vertical beaters at the back on both sides that would spray manure forty or fifty feet in the air and just as far to either side of the machine. This thing was designed to cover very large fields, so Rubin finished our new plot in about fifteen minutes. Then he planted it all in winter rye.

Finding workers wasn't quite as easy. Despite my tongue-in-cheek antipathy toward Eliot Coleman, he was still one of our most important sources of information, so we once again turned to him for advice. It seemed that almost all small organic farms relied on unpaid interns for at least some of their labour. The biggest agriculture school in the country is just an hour down the road at the University of Guelph, but at the time it had only one professor working on organics. We sent her an email asking if she had any graduating students who might want to intern on an organic farm. She wrote back and told us to join CRAFT—the Collaborative Regional Alliance for Farmer Training—one of those organizations that seemed to have chosen its acronym first and then thought up a name to go with it. CRAFT originated in Massachusetts, and a chapter had been operating in Ontario for a number of years.

CRAFT, we came to learn, was a grassroots organization in as close to the literal sense of that term as it's possible to get. It received no government grants or any other funding, it had no academic affiliations, it wasn't registered or incorporated anywhere. Legally, it didn't exist. It was just a bunch of small-scale organic farmers who offered full-season internships and who happened to live within reasonable driving distance of each other.

Gillian attended the last CRAFT training day of the season to get a feel for the group. She arrived to curious stares. It was a cold, wet day in October, and she was wearing pull-on leather boots, skinny jeans and a black leather jacket. Everyone else was in filthy work clothes and muddy rubber boots. Gillian hadn't realized that all CRAFT days include a group work project. She hadn't come prepared to get dirty, though she would have stuck out regardless of what she wore. Dreadlocks and beards abounded among the interns, and it would have been difficult to find a single facial feature that at least one member of the group had not managed to pierce.

It took a few hours before the other farmers realized that, despite her city clothes and the fact that she hadn't brought anything to contribute to the potluck lunch, Gillian was all right. They invited her to attend their winter planning meetings, and we were eventually welcomed into the fold based on what seemed to be an unspoken consensus among the farmers. What we didn't realize at the time was that we had unceremoniously joined a group of some of the most knowledgeable and experienced organic farmers in the country, many of whom had been at it for decades. They were also some of the most generous, grounded and lovely human beings I have ever met. We had already become members of two important communities—the first was our geographic community around Dunedin and Creemore, the second our food community of true-believer chefs—and now we joined a third, a community of farmers and apprentices who were quietly chipping away at the foundations of industrial agriculture.

We placed an ad for interns with a green-jobs website in December, and also sent a copy to the organic professor at Guelph. Figuring out the wording of the ad was a challenge. The only reason someone would perform hard physical labour on a farm for a full season for no money would be to learn something, but our qualifications as teachers were dubious. Then Gillian came up with an ingenious pitch: the best way to learn how to start a new farm is to work on a farm that's just starting! The amazing thing about this recruiting angle was that it worked. We received over two dozen applications for four positions within a week or two of posting the ad. The overwhelming majority of applicants were in their early twenties, university educated, raised in the suburbs and female. Exactly the kind of person everyone thinks of when picturing a farmer, right?

Anna was the first to come to the farm for an interview. She had grown up in Vancouver and had just finished a degree in international development at the University of Guelph, where she had taken a course on organic agriculture during her final year, just for the hell of it. I asked if she had any experience doing hard physical labour. She thought about it for a few seconds. "Not really," she said. "Although I did ride my bike from Calgary to Vancouver last summer. Does that count?" The fact that Anna wasn't sure if riding a bike over the Rockies qualified as "hard physical labour" impressed us. We hired her on the spot.

We conducted a lot of phone interviews over the next several weeks. Actually, it would be more accurate to say that we were the subject of many phone interviews. When you're offering prospective employees a salary of no money whatsoever, you don't get to ask a lot of tough questions. It eventually got to the point where if you were willing to intern and didn't profess to be a drug dealer or an arsonist, you were in. We ended up taking on four women, none of whom had any experience that remotely qualified them to work on a farm.

The final and most satisfying component of our expansion plan was mechanization. Tractors hold a special place in rural culture, and the kind of machine a farmer drives says a lot. Brand loyalty runs deep, often spanning several generations. It's not unlike English soccer fans; wearing the wrong tractor colour can get you beat up in certain situations. I received much advice from family and neighbours, all of it delivered with an air of grand authority, and all of it contradictory.

Tractor size is measured in horsepower: a tiny garden model might be fifteen horse, while the kind of multi-wheeled behemoths that tower over automobiles are three hundred or more. My father-in-law told me, "It doesn't matter what size you get, because you'll always want one size bigger." He managed to say this in a way that sounded sage. Rubin told me that I must get a tractor with a cab, so I could still get things done in bad weather. Judging by the two big blue Fords he owned, I understood "get things done in bad weather" to be some sort of Homer Simpson–esque code for air-conditioning, stereo and a fancy captain's chair designed to throw off an inquiring wife. He was also of the opinion that buying anything other than a Ford or John Deere risked a short slide into homosexuality.

I spent hours poring over used-equipment websites, leafing through the *AgDealer* and reading up on different brands. Then one day I saw what looked like a beauty—a fifty-horse Kubota with only two hundred hours on it, for $22,000. Tractor age is measured in hours rather than miles; two hundred hours is practically brand new. Kubotas are the Toyotas of tractors—better made, better resale value and more reliable than North American brands, and universally despised by my conventional neighbours. Kubotas are orange. No one around here would be caught dead on an orange tractor. They already thought I was a weirdo, so I bought the Kubota.

Thus began a deep and abiding love affair between man and machine that continues to this day. We try to use the tractor as little as possible, but that doesn't mean I love it any less. It's a wonderfully simple machine, easy to maintain, reliable and strong. I even learned how to change the oil, something I've never dreamed of attempting with a car. Despite what my father-in-law told me, I think it's just the right size. Tractors also require implements to do work—seeders, cultivators, snow blowers—the list is gloriously endless. I started with a six-foot-wide Rototiller that was delivered on the first warm day in the spring of 2009. A new season had begun.

THE INTERNS ARRIVED IN EARLY APRIL. Anna and Talia had come to the farm for interviews, but we had only talked to Melissa and Jenna on the phone, so we met them for the first time when they showed up at the start of the season. Many CRAFT farmers put up their interns in their house, essentially making their employees part of the family during the growing season. Sharing our home with four millennial women for six months sounded like our own personal version of hell, so we decided to rent a small house in Dunedin from our neighbour Donna Hamill. Donna and her husband, Wayne, are retired and live on our road, halfway between the farm and Dunedin, but they still own the house in the village where they raised their kids. Donna is like the godmother of the valley. You can often see her driving the two miles from her house to the village on her riding mower, on her way to cut the grass at the church. She was very reluctant to let four young women live in her house, but it had been empty for a while and she wanted the rent. She finally agreed after we promised to ensure that "gentlemen callers" did not become too numerous. She checked in at least twice a week, under the guise of mowing the lawn.

All winter I had dreamed about running the farm with a full crew of interns. I pictured strong young people accomplishing in a few hours what it would have taken Gillian and me days to do on our own. I would be able to sit inside at my desk on occasion, catching up on paperwork or making sales, while work continued outside in the field. This fantasy came crashing down on the very first day, as I stood staring at four twenty-somethings standing on my back deck, blank looks on their faces.

It was the first week of April. It was just a few degrees above freezing, very wet and windy. The snow was barely off the fields, and nothing was growing. I soon discovered that the stress of having way too much work, a stress that had dominated my life for two years, was nowhere near as bad as the stress of having a bunch of employees and nothing for them to do. Early spring is a time to build and repair things, to get ready for the growing season, but our interns didn't know how to build or repair. They didn't know how to do anything. Melissa had never even operated a hammer before. I had also grown accustomed to working alone and doing things at my own pace. That cold morning in April I didn't feel like working outside, but we had four interns ready to go at 8 A.M., and they would be ready to go at eight, six days a week, for the next six months. I panicked. "Today," I said, "we're picking rocks."

Rock picking is a task as old as agriculture (I just made that up, although it's probably true). It might also be the simplest manual task a person can perform. You see a rock in the field. You pick it up. You put it on the truck. Rubin had told me that the first rule of rock picking is "Only pick rocks that are bigger than your head." I'm a sucker for rules that use the head as a size reference, but Rubin's rule seemed appropriate only for large, mechanized farms. Even small rocks can interfere with hand tools, so we picked everything down to golf-ball size.

The six of us would each take a plastic bin, the kind you buy at the grocery store to replace disposable plastic bags, and crawl along through the five-acre field Rubin had opened up the previous fall, picking up rocks. When the bin was full, we dumped it into the back of the pickup. When the pickup was bottoming out on its springs, I would drive it over to the ditch at the side of the road and push all the rocks out. We did this every day for the first week, except one day when it snowed and I told everyone to stay home. After a few hours on the first day, when I thought the interns had mastered the art of rock picking, Gillian and I went inside to work on the computer for an hour. When I came back to check on their progress, I was met with unhappy faces. The interns made it clear that they expected either Gillian or me to work side by side with them, all day, every day. In their mind, that was the deal. They would work for no money, but they wouldn't work on their own. They wanted to learn. Apparently we were the teachers.

We picked rocks for two weeks straight, until it warmed up enough to start planting. In retrospect, it was probably the best way for the interns to start life on the farm. Over the years I have developed a theory about work—and by "work" I mean the kind of work we do on the farm: manual labour, often repetitive, that requires dexterity, strength and endurance. Most urban North Americans have never worked and don't know how to work. They are not physically competent. Our interns were no exception. Before I could teach them to do specific tasks on the farm, I had to teach them to work. They had to learn how to perform a physical task, to master the technique, and then do it over and over again, becoming faster and more efficient with repetition. Rock picking was the perfect starting point. It was simple but also physically demanding and very tedious. Over those first two weeks they became stronger and more competent and learned to cope with the mental challenges of repetitive manual labour.

Once the interns had learned how to work, new tasks became easier to master. Each new job was learned more quickly, even if it had no relation to rock picking or to any of the other tasks they had learned. We also did what people have done throughout history when faced with long hours of monotonous labour—we sang songs, we told stories, we talked and talked and talked. After two weeks of rock picking I knew more about Anna, Talia, Melissa and Jenna than I did about many of my closest friends.

OUR NEW, ROCK-FREE FIVE-ACRE FIELD seemed huge at the start of the season, but we quickly started filling it up. We planted beets, carrots, spinach, radishes and peas in the new garden, as well as much bigger weekly plantings of lettuce. We bought wider row cover so we could plant bigger beds of mustard greens, and we also covered rows of broccoli and cauliflower. We started almost a thousand tomato plants in the little green-house and moved them out onto the tomato fences in mid-May. We had laid out all the gardens so that all the rows would be a standard two hundred feet long. The new tractor-and-tiller com-bination prepared a six-by-two-hundred-foot swath for planting in about twenty seconds, a task that would have taken me at least half an hour with the walk-behind. And the tractor did a much better job.

Our radishes and arugula were the first things ready that spring. Gillian emailed Paul and Grace at 100km Foods on the Thursday of the third week in May and told them we would be ready for pickup on Monday. When the order came in on Sunday night, something had changed: our orders were up. Way up. Over the winter, some sort of critical mass had been reached, and it seemed that every serious chef in Toronto had suddenly been turned on to the idea of local, seasonal, ingredient-based cooking. Now it was spring, and everyone wanted our stuff.

As our other greens came on, our salad sales went through the roof. We scrambled to plant more and quickly fell into a rhythm of cutting greens for hours on end, several days a week. We were still working with the orange hand-cranked salad spinner that had saved our asses in Season Two, but now it could barely keep up with the volume we were putting out. It was also difficult for women to spin properly. The only way to get it spinning fast enough to dry the greens was to hug it to your chest with one arm while cranking the handle with the other, a configuration that women found uncomfortable in the extreme. As the only one without a pair of breasts, it fell to me to spin all the salad. My right arm became very strong.

One new client we picked up that spring started buying huge quantities of everything we offered. Reds Bistro was an eighty-seat place at the base of a bank tower in the centre of Toronto's financial district—absolutely prime restaurant real estate. Reds was corporate-owned, and a guy named Mike Steh ran the kitchen. We had met Mike at a few events, but we didn't know him very well. He had grown up in a working-class suburb of Toronto, and he didn't take any shit. His cooking was high-end, but he swore like a truck driver and had little time for niceties.

In late June, orders from Reds suddenly jumped. No restaurant had ever ordered so much from us in one shot. The next week we got an urgent message from Paul at 100km. "Mike Steh wants to set up a conference call to talk about price," he said. "He's pissed."

It turned out that Mike had gone on vacation and his sous-chef, Matt, who had become infatuated with our produce, had gone hog-wild with his ordering. Mike returned from his holiday to find his corporate bosses blowing their gaskets over a sudden spike in spending. Fine dining establishments like Reds usually limit food costs to about 28 to 30 percent of the menu price, so the food in a forty-dollar entrée should cost the restaurant no more than twelve bucks. Most of the rest goes to salaries and, in a place like Reds, to rent. Matt's unrestrained purchasing of New Farm produce had thrown off the economy of the whole restaurant.

I didn't like the idea of being berated about our prices by Mike Steh, but Gillian came up with some compelling reason as to why I should be the one to take the call. I was especially hesitant because we didn't have any real justification for what we charged. In a perfect world, farmers would base their prices on the cost of production plus a reasonable profit, but back then we had no idea what it cost us to produce any given crop. We knew that the farm was losing money, but that was about as deep as our financial analysis went.

I'm convinced that one of the biggest contributors to the perennial farm financial crisis is that most farmers harvest their crops before they are sold, then take what the market is offering. Some commodities can be stored for months or even years, which can allow a farmer to wait for better prices (if they have the money to survive while they wait), but fresh produce must be sold and eaten within a few days. A large proportion of the fresh produce that is consumed in eastern Canada passes through the Ontario Food Terminal in Toronto, a sprawling complex near the Gardiner Expressway where farmers, importers, distributors and retailers buy and sell fresh fruits and vegetables in massive quantities—it's the third largest food hub in North America, after Chicago and Los Angeles. A farmer who harvests a truckload of broccoli and takes it down to the food terminal has a powerful incentive to sell it as fast as possible. Every day the value of the produce drops, until it's inedible and worth nothing. It's easy to see why a grower might sell at a loss rather than bring home a truckload of rotten vegetables. Every player in the long chain from field to table is aware of the imperative of the fast sale, and every player uses it to squeeze the farmer.

Working with 100km Foods allowed us to avoid this trap by selling everything before it was harvested (though our farmers' market sales were a different story). Most of the money and labour required to produce our vegetables is in the picking, washing and packing stage. We don't even weed stuff unless we're pretty sure it's going to be sold. If we don't sell a bed of lettuce or a few rows of beets, we just till them under and return the nutrients to the soil. We lose the cost of the seed and the labour to plant it, but that's not much. Selling before harvest also helps us avoid uncomfortable discussions with chefs about money. We aren't showing up at the back door of a restaurant with a flat of tomatoes, then haggling over the price. If chefs think our stuff is too expensive, they don't have to order it. But this time Mike was angry, and we had to deal with him.

Paul called me, then conferenced in Mike, who launched straight into it. "Your stuff is killing me," he said. "It's great stuff, we love working with it, but it's way too expensive." He went on like this for some time. I think I'm a little older than Mike, but he addressed me in a tone that I imagined him using with a junior line chef who needed to be set straight. His rant ended with a blunt demand. "I'm paying over fifty bucks for a box of your salad. I can get the same organic mix from California for less than half that. You have to reduce your price."

That was too much. "The California mix is not the same as our salad," I snapped. "Ours looks and tastes better, it's fresher, it lasts way longer, and it has all kinds of varieties that they can't put in the California mix because they can't be harvested with their mechanical cutters." I was getting really fired up. "Has anyone ever come into your restaurant and said, 'Why should I pay fourteen bucks for a burger when I can get the same thing at McDonald's for three?' Of course they haven't, because your burger is better."

Fuck it, I thought. There's no way I'll ever be able to sell at a price that will make everyone happy, so I may as well double down. "If you really think our salad is the same as the cheaper stuff, then you should buy the cheaper stuff, because even if we cut our price by 30 percent, ours would still be more expensive and we would be out of business. We love working with you, and we really want to keep you as a client, but we aren't going to lower our price."

Mike grumbled a bit. He clearly wasn't happy, though he didn't cut us off right then and there. The call ended with no clear resolution.

Reds didn't order the next week, but the week after that they started up again. Matt had evidently been put on a short leash, but their orders increased steadily thereafter. By the end of the season, Reds was our biggest single restaurant account. Mike has run a bunch of different restaurants since then, and his chefs have always used our stuff. He's a good friend, a loyal customer and a great cook. He still complains that our stuff is expensive, but he knows it's worth it.

BY MID-SEASON, OUR MECHANIZATION STRATEGY was paying big dividends. The tractor purchase followed a pattern that would repeat itself for the next several years on the farm. Gillian and I would agonize over a big purchase, put it off for months or years, then finally knuckle under and spend the money (which we often didn't have), only to find that the thing we bought paid for itself in about six days. Then we'd spend some time beating ourselves up for not buying it sooner. The main problem with the tractor was that it was so efficient we were tempted to use it for everything.

Our philosophy from the beginning had been to limit off-farm inputs, especially fossil fuels. On many large farms, diesel fuel is the single biggest input cost. Farms running big machinery usually have large diesel tanks installed on the farm that are refilled by tanker truck. We weren't in any danger of going down that road, but we began to see "tractor creep" entering into our thinking. I once read that a barrel of crude oil can replace about forty thousand hours of manual labour. A statistic like that makes the rampant mechanization of agriculture seem almost inevitable.

We came up with a policy to use the tractor only for the things that would have the biggest impact on our efficiency, and to continue to do as much as possible by hand. That translated into using the tractor to plant and turn under our cover crops and to till the fields in preparation for planting. These tasks would have required an army of workers with hoes and shovels, or several teams of horses, equally unappealing options. Planting, weeding, harvesting, washing and packing would continue to be done by hand.

The only exception to this rule was potatoes. Our potato production had been limited in the first two years because potatoes require so much back-breaking labour to grow without machinery. So we bought a brand-new single-row planter for six hundred dollars that mounted on the three-point hitch at the back of the tractor. It had a seat where one person would sit and drop seed potatoes, one at a time, down a steel tube—the kind of mind-numbing all-day task that makes you pine for your old desk job. We also bought a single-row hiller to hump the potatoes. When I drove with the tractor straddling a row of potatoes, it would pile up soil against the plants from either side.

The first time I used the hiller, I didn't know what to expect. The potato plants were about a foot high, just tall enough that I could still clear them with the tractor. I didn't want to uproot a row or expose the developing tubers, which were like little white peas at that stage. I drove over the first row, lowered the hiller and drove forward a few feet. I raised the hiller, got off the tractor and adjusted the depth, then hopped back on and drove forward a few more feet. I couldn't stay centred on the potato row while at the same time looking back at the progress of the hiller, so I kept my eyes ahead and concentrated on staying straight. What the hell, I thought. I put the tractor into a modest gear and drove on.

I lifted the hiller as I drove out the end of the first row, turned the tractor and looked back. I could barely believe my eyes. The row was perfect—a uniform ridge of dark soil with about six inches of potato plant sticking out along the top, and not a weed to be seen. I headed back over the next row. Perfect as well. I realized that the hiller actually worked better the faster I went, so

I cranked the tractor up a couple of gears and hit the next row. I hilled the whole garden in about ten minutes, a task that would have taken at least a week with a hoe. As I sat contemplating my perfectly hilled potato patch, I began to cry. I just sat there on my tractor in the middle of the field, all by myself, weeping. I wept with sadness for the little bit of my life that I had lost to humping in the past. And I wept for joy at the thought that I would never hump again.

AT SOME POINT DURING SEASON THREE, it became clear that if our farm was to become successful, we would need to master a delicate and precise balancing act in our vegetable production. The amount we could plant was no longer a limiting factor. The combination of the tractor and our old EarthWay seeder allowed us to plant large areas in a relatively short period of time. But it was of no use to plant three acres of salad greens if we didn't have enough people to weed and harvest that much, let alone try to sell it. We needed to plant the right amount at the right time, so that the six of us would be kept consistently busy every day, while meeting the demand from our customers and maintaining a high level of quality. The challenge was that two of our biggest variables—weather and sales—were in constant flux. This led us to err on the side of caution and plant significantly more each week than we thought we would sell, which in turn led to all of us being overworked and stressed out, which led me to become, at times, an asshole.

On one particularly hectic harvest day we were scrambling from six in the morning, trying to cut and wash everything in time to meet the 100km truck. We tied down a precarious stack of bins on the back of the pickup, then I jumped in for the forty-five-minute run to Barrie Hill. I stuck my head out the window and yelled a few instructions to the interns: "We need two beds of greens planted next to the stuff we planted last week, but first put in a bed of salad turnips over there." I pointed to the far end of the garden. Then I raced down the lane.

Salad turnips are a weird vegetable. They look like a big radish but they're sweet and spicy tasting and have the juicy consistency of a Fuji apple. The variety we grew back then was called Hakurei. It was pure white, inside and out, perfectly round, and could grow to the size of a softball. Salad turnips have very, very small seeds.

When I got back from Barrie Hill, Gillian was at work in the office and there was an intern conference going on in the garden. This was never a good sign. Intern conferences usually led to intern delegations, which usually resulted in a prolonged discussion of feelings. I don't know if all twenty-somethings have a lot of feelings, but the twenty-something women who were working on our farm that year sure did. It wasn't enough to talk about the work we had to do or to discuss a problem with the intern living arrangements—we also had to talk about how we felt about those things. I'm not a big feelings guy, so those discussions made me sad.

I walked out to the field and asked what was going on.

"We're out of salad turnip seed," Talia said.

"There's more in the seed cupboard in the mudroom," I replied.

"We already used that," she said. "There's no more."

It took me a few seconds to process this information. "There must be more," I said. "We have enough for the whole season."

The interns all looked at each other uncomfortably. "I thought maybe we had the wrong seed plate," said Melissa. I felt my ears start to heat up. You will recall that the EarthWay seeder uses interchangeable seed plates to plant different vegetables. The seed plate for corn has big scoops to pick up the big corn seeds. The seed plate for carrots has tiny scoops to pick up the tiny carrot seeds. There is no salad turnip seed plate—I had told them to use the carrot plate for salad turnips, but they had forgotten. They had decided that they should use the radish plate, because salad turnips are a lot like radishes. The only problem is that a radish seed is about twenty times bigger than a salad turnip seed.

It gradually dawned on me that the interns had dumped a season's worth of very expensive salad turnip seeds into four or five rows in the garden. What's worse, they had all gotten together and consulted with each other before doing it. This made me angry. This was stupidity by committee. I lost my shit. I was tired and overworked and stressed out, and I went crazy. I don't recall exactly what I said, but my F-bomb-per-sentence count was impressively high, and I improvised some arm gestures and facial expressions that I think paired well with the expletives. I stormed off in a cloud of vitriol.

Gillian came and found me in another part of the garden about half an hour later. There was another intern conference going on, this time in the wash shed. Gillian explained that the interns were very upset, that my harsh words and angry tone had hurt their feelings, and that at least two of them were in tears. "Good," I snapped. Gillian remained calm. "You need to apologize," she said. I almost lost it again. Me, apologize? They should be the ones to apologize! But I knew I wouldn't win this battle. Farm workers can be unhappy or unpaid, but they can't be both.

I found the interns huddled in the wash shed. Their eyes were red, and they looked a bit as if they had just witnessed a terrible car accident. "I'm sorry," I mumbled. "I lost my temper, and the way I acted was wrong."

"The way you acted was wrong!" said Anna, suppressing tears.

I was about to point out that I had just said the same thing, but quickly thought better of it. I realized that there was to be no admission of guilt from the interns. They may have made the initial mistake, but I had committed the ultimate crime: I had lost my temper. Such a crime was, I knew, punishable by discussion, so I stood there and discussed, calmly accepting their repeated criticisms, silently wishing I were somewhere else.

THE SECOND WEDNESDAY OF EVERY MONTH is CRAFT day. On CRAFT days, the farmers and interns from every CRAFT farm in southwestern Ontario gather at one member farm for a panel discussion (about soil or marketing or greenhouse production, or something like that) followed by a farm tour, a potluck lunch and a work bee in the afternoon. There were sometimes upwards of sixty people at CRAFT days that season, representing ten or twelve member farms. It was a pain in the ass to shut down production in the middle of the work week and drive for two or three hours to get to the host farm, but it was always great when we got there. Everyone got to socialize, and the lunch, though sometimes heavy on the kale, featured some truly exceptional produce.

All the farms we visited had been in business for much longer than we had, but I was often underwhelmed by what I saw. I got the feeling that the farmers were still struggling financially, even after many years of work. I saw lots of weedy gardens and crappy old equipment, and I saw farmers making simple mistakes that even I could recognize as such. On the one hand, it made me feel that our farm wasn't so bad after all. On the other, it made me wonder if we were destined to continue scraping by for years to come. Then, on the second Wednesday of July, we had a CRAFT day at Meeting Place.

Meeting Place Organic Farm is run by Tony and Fran McQuail, American Quakers who came to Canada during the Vietnam era, part of an exodus of pacifists, intellectuals and artists that has enriched Canadian society in profound and underappreciated ways for more than forty years. When the McQuails bought their hundred acres outside Lucknow in the early 1970s, it had no house or barn. They started farming conventionally but were horrified by the toxic substances they were told they should use. They switched to organic when the term had just been invented, and they've been at it ever since, part of that small, incredibly important generation of farmers who straddle the back-to-the-land and the good-food movements.

Tony looks exactly as you would expect an organic farmer to look. Tall and lean, he has a white beard and long, snow-white hair that he keeps in a ponytail. He usually wears a straw hat. Fran is short and slightly cherubic; I often have the urge to hug her. They run holistic management workshops in the off-season, teaching farmers to view financial success as a means of achieving a sustainable and meaningful life, rather than an end in itself. Gillian attended one of Tony and Fran's sessions a few years after we first met them; they asked participants to write their own three-line eulogy, summarizing what they hoped their children, spouse and friends would say about them after their death. Gillian glanced at the stranger sitting next to her, who had written "He was a good man" three times, tears silently streaming down his face. Tony and Fran seem to have figured out what is truly important, and they have a way of helping others figure it out, too.

As we walked around their farm on that CRAFT day in 2009, I was struck by how different the place felt compared to our farm. Our farm was chaotic and messy and half-baked. Meeting Place was serene and tidy and complete. Ours felt manic. This place felt calm.

Tony first took us to see their pasture. The McQuails practised something they called "holistic management planned grazing" to produce their beef, lamb and chicken, a technique that involves moving animals frequently between small sections of pasture created with movable electric fencing. Grass grows slowly after being grazed, then accelerates rapidly after a few weeks. Traditional grazing systems allow animals to continuously eat the newest shoots, keeping the entire pasture growing at the slowest rate. Tony and Fran allow their cattle onto a section of pasture only when it reaches the top of the productivity curve, which dramatically increases the amount of meat they can produce on their small farm.

When we walked into a new section of pasture with the cattle that day, it was like stepping into a waist-high salad bar. The field was lush and green and incredibly diverse. Tony showed us how he added alfalfa, trefoil and white clover seed to his mineral licks, so the cattle would literally reseed the pasture with their shit. After the cows had grazed down a small section for a few days, Tony would bring in the sheep to work over the less desirable grasses that the cattle had left. Meat chickens in movable pens came through last, eating more grass, plus the bugs that hatched out of the cow and sheep droppings. Then the section would be left to regenerate for several weeks before starting the cycle again. It would be a lot less work to simply let the animals graze over the entire farm for the whole season, but that would have made it impossible to make a living on a hundred acres.

Fran's vegetable garden was equally impressive—orderly and weed-free. Beside the vegetables was the horse paddock. Some of the work in the hay and grain fields at Meeting Place was done by tractor, but everything in the vegetable garden, and a lot more besides, was done with horses. The McQuails bred and worked Belgian draft horses. On the day we were there they were running back and forth in the sun, their blond manes streaming in the wind like an equine reincarnation of ABBA.

Everything we saw was inspiring, but for some reason, what has stayed with me from that day is the apple orchard. When Tony and Fran first bought their farm, it had an old, overgrown orchard of McIntosh trees. McIntoshes are notoriously susceptible to scab, a fungal infection that makes the apples look scabby (no surprise) and causes the leaves to wither and turn yellow, then brown, then drop off. When the McQuails began rehabilitating the old orchard, they were advised to spray with fungicides, which they did. They found that they had to spray over and over again to keep the scab in check, which is standard procedure for conventional apple growers. After a few seasons of more or less continual spraying, they found that scab was still a problem, so they bought a new fungicide called Captan. Before they could spray, the news broke that the manufacturer had falsified the results of the toxicity studies. That was enough. "I walked into the farm supply store," said Tony, "and I returned the bags of Captan." That season, the scab took over the orchard, almost completely denuding all the McIntosh trees. The apples were covered in black scabs and were about the size of a quarter

Tony resolved to cut down all the trees and replace them with a more scab-resistant variety. But as fate would have it, when Tony went to do the job late that fall, his chain saw broke. It was winter by the time he got it fixed, so he decided to wait until spring. When spring came, things got busy on the farm, as things tend to do, and Tony never got around to cutting down all the trees. When fall came, Tony and Fran found to their surprise that the apples were not quite as bad. There was a lot of scab, but it was better than the year before. Each season after that saw improvement, with less and less scab and better and better apples.

Tony did some research to try to find out what was going on in his orchard. He found that the spores of the scab fungus overwinter on fallen leaves under the apple trees. Dead leaves are broken down and incorporated into the soil primarily by earthworms, but fungicides are toxic to earthworms. The more the

McQuails sprayed, the fewer earthworms survived in the orchard and the more scab-infected leaves there were the following season to reinfect the apple trees. Tony inadvertently broke this cycle when he stopped spraying. In later years he found that scab outbreaks tended to be small and short as long as he *didn't* spray. The earthworms that thrived in the pesticide-free soil under the trees quickly ate up the scabby leaves as they fell, limiting any outbreak. Tony aptly called this strategy "benign neglect" and said it came to shape his view of everything on the farm. Mother Nature has a way of finding balance. Often, you just need to get out of the way and let her do her job.

Gillian and I followed Tony and Fran all over the farm, not wanting the day to end. We were in the presence of deep knowledge and profound wisdom, possessed by generous people who were happy to share all they knew with other farmers who could easily be viewed as competitors. I couldn't get my questions out fast enough. Tony and Fran had built everything with their own hands—their house, their barn, their farm, their business—but more importantly, they had built a satisfying, meaningful life for themselves and their family. Here was proof, I thought. It can be done.

AS SEASON THREE BEGAN TO WIND DOWN, I stopped to marvel at the changes that had come over our farm. We had more than doubled our garden area and our sales, built a walk-in cooler and covered outdoor wash area attached to the wash house, put up a new hoop house, and dramatically increased our production capacity by semi-mechanizing our pre-industrial growing system. We set a new one-day record for sales in the farmers' market—over two thousand dollars—and had to take two vehicles to Barrie Hill for the 100km drop because our orders were now too big to fit in the pickup. The business was still bleeding money, but it was now more of a slow drip than a life-threatening hemorrhage.

An amazing change had come over the interns as well. They were physically strong, lean and healthy, and they had also developed a great deal of mental toughness. They seemed more confident, more mature and maybe even a little more serious. Melissa had been a giggly incessant talker when she arrived; by the end of the season she was still unfailingly positive and happy, but she no longer felt the need to fill every conversational void. Gillian and I looked at those women with a great deal of admiration, and not a little pride. They could do just about any task on the farm with skill and endurance; they had learned not only how to swing a hammer but also how to fire a gun, drive a tractor and plumb a wall, among many other skills. Those women really knew how to work.

The final task of that season, in late October, was to kill the pigs. We had learned a few things from the fiasco of the previous year. We avoided the whole castration problem by purchasing two female piglets. Anna, Talia and I had picked them up in the spring and released them into their pen without incident or escape. They had grown up over the summer, happy and contented, with no humping or disturbing sexual displays of any kind, and—best of all—no testicles. We also decided to avoid the trauma and cruelty of transporting them to a commercial abattoir by killing and butchering them on the farm.

One of our first and most supportive chef customers was Mark Cutrara. Mark had his own restaurant back then, a very small place called Cowbell on Queen West in Toronto, in a neighbourhood that was just starting to gentrify but was still pretty rough. He had Lennon glasses and a Lenin goatee, and a quiet, introspective demeanour that is not common among chefs. Mark had experienced an epiphany while eating a meal of foraged wild food on the coast of British Columbia (what he called "the greatest meal of my life") and had returned to Toronto with an obsessive devotion to the best seasonal ingredients. He practised "nose-to-tail" cooking, bringing only whole animals into his tiny kitchen and using every part of them, doing everything in-house, right down to churning his own butter. His menu was a chalkboard propped in the corner of the room that changed every day.

When we decided to kill the pigs ourselves, Mark was the first person we called. He was eager to help. Mark and his crew, like all good nose-to-tail chefs, were experienced and skilled butchers, but none of them had ever killed an animal before. Mark suggested that they come up for the kill on a Friday, then return for the butchering on Sunday. That would give the carcasses time to cool. Butchering an animal while it's still warm and floppy is difficult, and it greatly increases the odds of inadvertently cutting off one of your own digits.

I decided it would be wise to bring in a professional for the kill, since no one on either the Cowbell or the New Farm team had any direct experience. I asked Rubin if he knew anyone who could help and, of course, he did. He gave me the number of Frank the Butcher, an itinerant killer of animals who lived somewhere over in Grey County. Frank's number had been disconnected when I tried to call him, but Rubin got a message to him and assured me that he would show up on the appointed day.

The third Friday in October dawned cold and grey and drizzly. Mark and two of his chefs arrived just after eight. Frank showed up a few minutes later in a very large and very beat-up pickup. He was extremely rough looking, dressed in blood-stained canvas coveralls and with about a quarter of an inch of stubble. When he spoke, he seemed to leave off the last half of most of his words, so it was difficult to understand what he was saying.

I had read quite a bit about killing animals on the farm and had laid out a small arsenal for the job: the .22 rifle for the kill shot, my .30-06, in case the .22 didn't hit the brain and the pig started running around, and my 12-gauge, in case Frank thought a shotgun was a better back-up piece than the .30-06. A pig's brain is relatively small and they have a thick skull, so if you don't shoot it in just the right place the first time, you end up with an angry, mobile pig on your hands. My absolute favourite illustration of all time, from *The Encyclopedia of Country Living*, shows exactly where you want to aim:

Shoot here

Frank suggested I shoot the first pig and he would stick it. A bullet to a pig's brain will cause the pig to drop and lie motionless for a few seconds. The dead pig's nervous system will then kick in, and it will flop around lifelessly for a minute or two before it at last just lies there like it's dead (which it has been from the moment it was shot). It's important to stick the pig during the initial seconds of calm so that it bleeds out as much as possible. The encyclopedia suggested sticking the pig in the neck, to cut the jugular veins. Frank said he preferred to stick a pig directly in the heart.

I hadn't fed the pigs the night before, to make them easier to gut, and they were hungry. We opened the gate between the pigpen and the chicken yard and I presented the closest pig with a rubber bowl full of food. It trotted through the gate eagerly and nuzzled my leg as I walked it toward the middle of the enclosure. I put down the bowl and it went right at the feed. I picked up the .22 and put the muzzle against its forehead, just like in the diagram. It was moving a little and my heart was beating like crazy, and I hesitated. My hands were sweating. The anticipation of killing this pig was completely different from the feeling I had had the previous fall, when I was about to shoot my first deer. My physical reaction was the same—especially the rapid, pounding heartbeat—but with the deer I felt an elated, primal adrenaline rush, as if I was finally doing something that I was always meant to do. Here I felt sick to my stomach. Killing the pig felt cynical, calculated and completely unnatural. I knew this animal. I had seen it display emotions and maybe even empathy. I had taken care of it and provided for it, and now I was going to take its life. Killing the pigs on the farm was supposed to be more humane then taking them to the abattoir, as we had done the year before, but now it was the killing itself—regardless of where or how it was done—that seemed inhumane.

The pig tossed its head, annoyed at the gun muzzle hovering just above its eyes. I gritted my teeth, lowered the barrel and quickly pulled the trigger. The .22 let out a loud crack. The pig went completely rigid for a split second, as if it had received an electric shock, then collapsed and keeled over on one side. Frank was moving before it hit the ground, a long, thin knife with a slightly curved blade in his hand. He placed the tip of the blade just to one side of the pig's sternum, high up on its chest, and slid it in to the hilt and back out again in one fluid, seemingly effortless motion. A fat ribbon of dark red blood spurted out of the wound in a clean arc. Frank stepped back and the pig began to twitch, then its legs began to move spastically, kicking at the air. Blood erupted from the pig's chest in short spurts. After a minute or two, the movement slowed and then stopped.

We were a strange-looking group standing silently in a circle around the dead pig, steam rising from the blood-soaked grass: a couple of hipster cooks, a few earnest interns (one a vegetarian), a redneck butcher and me. Gillian was in the house. She had seen it all before when she was a young girl on the farm in Vermont, and she didn't care to see it again.

I brought the tractor into the yard. Frank cut a slit between the tendon and the bone of the back legs and put a steel rod with a hook through the slits. We then hooked the rod to the tractor bucket and lifted the pig into the air, head down. Pigs are usually scalded immediately after being killed, which means they are immersed in very hot water to loosen their hair, which is then scraped off. I had wanted to scald the pigs, so I had gotten up early and started a fire to heat two big steel garbage cans of water. By the time the first pig was up on the hoist, though, the water still wasn't hot enough, so Frank decided to skin it.

Skinning a pig is very difficult. Their skin is thin and breaks easily, and there is a thick layer of fat underneath that's an important part of many cuts of meat. That's why scalding is the preferred method—the skin is the delicious rind on a pork roast that holds in the equally delicious layer of fat. Frank made it look easy. He cut the skin off the pig with almost no fat on it, and nearly unbroken. Next he cut a ring around the pig's asshole and tied off the anus. Then he made an incision in the pig's crotch, reached in and pulled the anus through to the abdominal cavity. He put his fist into the cavity with his knife blade facing out, and with one smooth motion, sliced all the way down to the sternum. His knife was so sharp and he was so strong that he cut smoothly through the breastbone, all the way to the pig's throat. All the guts, from colon to windpipe, fell out in one connected mass, revealing a pure white, empty cavity. The heart had a clean slit right in the centre, where Frank had stuck it.

The second pig had been just over the fence while all this was going on, blissfully unaware of what was befalling its sister. It, too, trotted eagerly into the killing pen. It, too, collapsed instantly when shot. This time Mark did the sticking, under Frank's careful instruction. He made a perfect stab to the heart but didn't pull out fast enough, and his whole arm was soaked by the blood that gushed from the wound. I saw Mark standing off to the side as the pig writhed, observing the steam rise from his bloody arm with an expression of morbid fascination on his face.

Two days later, Mark and the Cowbell crew returned to butcher. Over the course of several hours, two whole pigs were dismembered, processed and packaged. Quarters were reduced to primary cuts, which were in turn reduced to the parts of animals that we all recognize—chops, roasts, steaks. The bellies and hams were put into a large bin full of water, salt and spices, and left to cure. The trim and some of the meat from the shoulders was ground up and stuffed into casings to make sausage. Mark had recipes for everything, and we used garlic, onions, shallots and herbs from the farm to flavour the sausages and hams. The interns had no experience with any kind of knife work, but after six months on the farm they knew how to work, so they weren't intimidated, and they picked it up quickly. Gillian did double duty, butchering in the morning, then making a spectacular farm lunch of roast chicken and vegetables that we ate outside, despite the fact that it was barely above freezing. By the afternoon everything was wrapped in butcher paper, labelled and in the freezer. The season was over.

CHAPTER 9

CASH CROPPING

SEASON FOUR WAS THE FIRST to start with a sense of optimism rather than impending doom. We had broken a hundred thousand dollars in sales in Season Three and had been unable to keep up with demand. Gillian was still forced to do some consulting work over the winter to make ends meet, but at least we had a glimmer of hope that we would soon turn a profit. For the first time, our goal of creating a farm that could support our family seemed like it might be attainable.

It was obvious that our interns were the key. Bringing on outside labour had radically increased our productive capacity and had allowed us to grow enough food to reliably serve more high-volume customers. Now it was 2010, and it was clear that the local-food movement was far more than just a fad. Demand was growing, more and more chefs were climbing on the farm-to-table bandwagon, and Paul and Grace had ambitious expansion plans for 100km Foods. We decided to go big.

We did a careful analysis of our production systems over the winter to identify bottlenecks and potential efficiencies, and zeroed in on an unlikely target: the orange salad spinner. Our greatest labour demand was on harvest days, and the most obvious bottleneck in our harvest process was drying salad. So we bought a big self-contained electric salad spinner. It was basically a stainless steel drum on wheels that stood about three feet high, with an electric motor on the bottom and a big decal on the side that read "Salad Ace." It looked a lot like R2-D2, minus all the dials and flashing lights. The Salad Ace cost more than $2,500, but it dried salad three times as fast as the orange hand-cranked model, with a lot less effort. We also tripled our cucumber-growing capacity with a brand-new hoop house that was taller and wider than our original secondhand structure. Then we got more interns.

Our friends in Dunedin always gave us a hard time about our interns. "I can't believe you convinced all those girls to work for you for free!" they would often say. To be honest, I couldn't believe it either, until I realized that they weren't working for free at all. They may have been unpaid, but they were far from free.

For starters, we paid each intern a stipend of fifty dollars a week to help cover the cost of food and other living expenses. It wasn't much, but it added up.

More importantly, as we learned in Season Three, the interns were loath to work alone. Small-business owners often make the mistake of spending too much time working *for* their business and not enough time working *on* their business. Being forced to work alongside the interns all day ensured that we were always working *for* the production side of business rather than working *on* the sales, marketing and planning that were the key to profitability.

Keeping the interns happy also cost a significant amount of money. There were lots of days when we had to shut down production and leave the farm for intern activities—CRAFT days, field trips to other farms. It was a guaranteed morale booster when we took all the guns to the back of the farm and spent half a day shooting stuff, but we certainly didn't make any money on those days.

We also had to provide the interns with housing. In Season Three, that had meant renting Donna and Wayne's place in Dunedin, but with our expansion plans, we couldn't fit all the interns we wanted into that little house. We talked to the township about building a bunkhouse on the farm, but the building code requirements would have made that project extremely expensive. We ended up buying a derelict one-room schoolhouse a half mile down the road at the four corners of Maple Valley, and renovating it into a three-bedroom cottage. Buying something off-farm had several advantages—it gave us an asset we could unload if the farm went bust, it gave the interns and us a bit more personal space, and we could rent it to skiers in the winter. But it was expensive—all in, it cost over two hundred thousand dollars to buy and renovate.

Training, however, was the biggest expense of all. When the Season Four interns showed up in April that year, they were just as clueless and unable to work as the previous lot. We had to start from scratch again, teaching everyone the most rudimentary skills and walking them through every step of our production processes, again and again. At one point early in the season someone made a comment in the market about our free labour. "Interns aren't free labour," Gillian snapped. "Interns are a bunch of people who need to be disabused of the idea that they want to be farmers." That was funny, but not entirely true. Our internship arrangement was a trade—labour for knowledge and experience—that we did our best to ensure was fair for both sides. But having spent half the winter recruiting, all spring building their housing and half the summer training, the word "free" made us bristle.

The final big investment that spring didn't do anything to boost production or increase sales, but it helped move us toward one of the larger goals of the farm. We installed three large arrays of solar panels in the field beyond the greenhouses. The power was all sold into the grid under a twenty-year subsidized contract, but the panels produced enough electricity to offset just about everything we used on the farm. Now we were more or less self-sufficient in electricity. The solar panels and the interns' schoolhouse forced us to borrow money for the business for the first time, but we had enough confidence in the farm to take that leap. We were no longer quite so hamstrung by a lack of equipment and labour. It was time to see what the farm could really do.

WE STARTED PLANTING IN ANOTHER new five-acre garden that we had prepared the previous year. We had again brought in composted manure from one of Rubin's farms to get the new ground going, but we didn't want to rely on off-farm fertility over the long term, so we continued to work on our cover-cropping regime. We decided to adopt a two-year cycle—each section of garden would spend one year growing vegetables, then the next year in cover crop. We would plant and replant in the vegetable year, growing as many as four separate crops, one after the other. Then, after the last crop of the season had been harvested, we would plant the section down in winter rye. The rye would be turned under in late May or early June the following year, when it was waist high, adding many tons of organic matter to the soil. In late July we would plant a cocktail crop to add even more organic matter, as well as nitrogen. We mixed various combinations of oats, barley, wheat, climbing forage peas, sunflowers, sorghum-sudangrass, alfalfa, crimson clover and hairy vetch. We even tried something called forage radishes, which grew enormous tubers that burrowed six feet down into the subsoil, bringing copious nutrients up to the surface. The varieties in the cocktail-crop mix were selected for their inability to survive a hard frost. They all died over the winter, creating a thick mulch of plant material that protected the soil and broke down quickly in the spring. The garden would then be ready for vegetables again, a year and a half after the last crops had been harvested.

By Season Four we had a total of thirteen acres of garden in play, which was pushing up against the maximum tolerated by adherents of the small farm orthodoxy. We planted vegetables in the new five-acre field, as well as in the three-acre section we had used in Season Two. The five-acre section we had used the previous year, in Season Three, was in cover crop. We found that the old Season Two garden was richer and more productive than it had been the first time we used it, which is what we had hoped. Our cover crops were putting more into the soil than the vegetables took out. It was the start of a virtuous cycle that saw the quality and productive capacity of our soil increase every year. The only inputs the system required were our labour, some seed and diesel fuel to run the tractor. Everything else came from the soil, the air and the sun, and didn't cost us a dime.

The interns arrived in dribs and drabs that spring—another hazard of relying on unpaid workers is that it's hard to insist on a firm start date. The first batch showed up in early April. Sarah set a new benchmark for knowing nothing about plants or how they grew. Ariella flew in all the way from British Columbia and brought a west coast vibe with her. I had instituted an informal affirmative action program that winter, in an attempt to finally recruit a male intern. It paid off in the form of Will Hill, a skinny, profane, incredibly funny guy from a small town in northern Ontario who was partway through a master's degree in food policy. I liked him right away, and Foster started following him all over the farm. The first group also included Katie, who was blond and very tall and had a kind of goofy, self-deprecating enthusiasm that was strangely endearing. She had grown up in the suburbs and graduated from Queen's University, then suddenly bailed on her corporate job and signed up with us.

The rest of that season was a revolving door of interns. Mike showed up in June. Then Ariella crashed her bike on our sideroad and badly injured her leg. After a month of my driving her to medical appointments (at which I was often humiliatingly asked if I was her father), she had to fly home. Mike's girlfriend, Annie, showed up for a visit at about the same time and ended up staying for the rest of the season. Rachel started in May but had to go back to school in September. Kirsten was our youngest intern—she started in late June when she graduated from high school, but sometime in September she went home for the weekend and never came back. Todd came on for two months at the end of the season, to replace Rachel. I can't remember when Susan started. We had hoped to have six interns for the whole season, but our numbers fluctuated between five and eight. Unlike many other farms, we didn't make meals for our interns, but Gillian and I still felt at times as if we were running a youth hostel.

We continued to refine our planting systems while at the same time trying to mould our collection of vagabond interns into an effective farm crew. Our perpetual battle with the flea beetles had by that time compelled us to buy new, heavy-duty row cover. This stuff was a polypropylene mesh rather than a woven fabric, and it never let in a single bug. It was twelve feet wide, and we figured out a way to fit twenty rows of salad greens under one strip—four thousand feet of row in total under one piece of row cover. We had to put more than a hundred plastic hoops in the ground to hold up each piece, and bury all the edges in dirt with large hoes. It took one person about half an hour to plant one bed, then our entire crew of eight or nine people half an hour to cover it.

We also continued to whittle down the number of varieties we grew, jettisoning more stupid vegetables that were hard to grow, hard to sell—or both. We found that no matter what we did, our cauliflower never tasted good (one of the Season Three interns said it tasted like cigarette butts), so we stopped growing it. We also got rid of garlic, all onions, winter squash, cabbage, head lettuce and cucumber varieties other than Tasty Jade, which clearly tasted the best. We completely gave up on selling anything perennial, because we couldn't keep up with the weeds; our asparagus, strawberries and raspberries were henceforth for home consumption only. That year we established a formal home garden, where we grew all the stupid vegetables that we liked to eat but no longer grew commercially. It was fun to grow things like celery root, cilantro, fennel and ground cherries without having to worry about selling them. The home garden fed us and the interns during the summer and helped fill our root cellar in the fall.

In July we hosted a CRAFT day for the first time, and Gillian and I gave a talk about marketing to about seventy-five interns and farmers. I told the crowd that they shouldn't feel compelled to grow hundreds of different varieties, that some vegetables simply don't grow well on some farms, and some are easier to sell than others. I of course used kohlrabi as an example—the stupidest of stupid vegetables. A farmer named Gavin put up his hand.

"I just want to clarify something," he said. "What you're saying is that *you* don't like growing kohlrabi, but it could be a good crop for other people. There's nothing inherently wrong with growing kohlrabi."

"No," I said quickly. "I'm saying that kohlrabi is stupid, and no one should grow it."

Everyone laughed, because it was true.

BY THE MIDDLE OF THE SUMMER, the Season Four interns had learned to work, and the crew was operating like a well-oiled machine. I watched in amazement one day while Sarah, who had arrived at the farm with a vague notion that beets might grow on trees, rapidly buried a section of row cover using the toes of her bare feet to stretch and position the edge of the fabric while simultaneously operating the hoe with her hands. It reminded me vividly of the dexterity and competence that I had seen women display while working in the fields in Malawi. We sold more than ever in the farmers' market, and our restaurant orders grew so large that Paul began bringing the 100km truck straight to the farm for pickup, so we no longer had to make the drive to Barrie Hill.

Our farm also began to attract a fair bit of attention. The year before, I had written an article about leaving the city and starting the farm that had appeared in *Toronto Life*. The magazine had sent a pair of photographers to the farm who followed us to the farmers' market, holding big reflector screens above us while we served our customers, taking millions of pictures and generally causing us a great deal of embarrassment. Our friend Dan saw the photographers hovering around our stand. "What's up with the farmerazzi?" he asked casually.

The *Toronto Life* piece led to more publicity. In 2010 it seemed like we had media at the farm every other week. There were magazine articles, radio pieces, the odd TV crew, and a feature in the *National Post* newspaper. Our friend Gord started calling us the *news* farm. There was a growing fascination with the idea that young, educated people might leave the city to start a farm. Most of the coverage had a strong *Green Acres* slant to it; the interesting angle was always the juxtaposition of the sophisticated urbanite taking on the backward occupation of farming. One journalist described us as "latte-sipping oenophiles," which was strange, since neither of us drinks coffee, and I had to look up the word "oenophile" (it means wine lover). It seemed that urban society viewed us in much the same way that our rural neighbours did—amusing, a little bit crazy and probably destined to fail.

As our very modest fame grew, the Toronto restaurants we dealt with started putting our name on their menus. That diners would care about the source of the food in a restaurant was a relatively new idea in 2010. One day my mom called to tell me that she had seen my name in the newspaper. It turned out that a place called Marben had taken the "farm on the menu" idea one step further and named all their dishes after the farmers who produced the ingredients. The *Globe and Mail* review my mom had spotted talked glowingly about two of Marben's salads—Gil's Greens and Brent's Beets.

The kitchen at Marben had just been taken over by a pair of young chefs who had met at the Stratford Chef School and had most recently worked with Mark Cutrara at Cowbell. Carl Heinrich was a bit of a prodigy. He was only twenty-three when we first met him, but he had already worked for years in some of the top kitchens in the world, including a stint with Daniel Boulud in New York. One reviewer called him "the Doogie Howser of the Toronto culinary scene." Ryan Donovan was a little older and was also a trained butcher. Carl and Ryan had taken the whole-animal, farm-to-table ethos of Cowbell with them to Marben, and the place was booming. They quickly became one of our biggest restaurant accounts.

ALL THE INTEREST FROM THE PRESS and restaurants translated into ever higher sales, but not the kind of breakthrough we had been hoping for. We had started the season feeling very optimistic and had pushed our interns as hard as we dared, but our sales growth was only modest. We ended up with just under $150,000 in sales for Season Four, which was enough to cover all our expenses and prevent us from going further into debt. We even managed to pay ourselves a little, but somehow that didn't make us feel a whole lot better. The fact that we had both worked like dogs for four years and had so far paid ourselves less than ten thousand dollars was profoundly depressing. I had always imagined a big celebration when we finally broke even, but we didn't feel much like celebrating that year.

We ended the season with our second on-farm pig kill. Gerald te Velde didn't have any piglets for sale that spring, so we had bought two from the Stadtländers. Michael told us that he never castrated his pigs and never had a problem with boar taint, so we got one male and one female. Carl, Ryan and the crew from Marben came to help that year, and we felt confident enough to do the deed without the assistance of Frank the Butcher. Everything went smoothly; we even got the water hot enough to successfully scald the pigs and retain the skin. Carl suggested we cook up a tenderloin on the fire immediately after we finished gutting the pigs. The tenderloins are the only cuts that are inside the rib cage, along either side of the spine, so they're easy to cut off a whole carcass. Ryan cut one out of the male pig. It was still warm, so it cooked in just a few minutes. It was hard to tell, out in the wind, with all the woodsmoke, but I thought it smelled a little funny while it was cooking. Carl cut off a piece while it sizzled in the pan and popped it in his mouth. "Yep," he said. "That's what boar taint tastes like."

THE PLAN FOR SEASON FIVE was much the same as it had been for Season Four. We couldn't really increase the number of interns because we didn't have space to put them. We had bought an old fifth-wheel camper and parked it behind the schoolhouse to give us a fourth bedroom, but we couldn't sleep more than eight people and we already had to use two cars to get the crew into town on days off. The imperative of working alongside our interns also made a larger crew impractical; there are only so many jobs you can do with ten people. We opened a little bit of new land that season, but without a larger labour force to work it, our ability to expand was limited.

We added some expensive new infrastructure—another hoop house, a second salad spinner, a one-row potato harvester and a big drum washer for our root vegetables—but we immediately ran up against the same reality that we had run up against the previous two years: when you're operating a human-powered farm, the skill and productivity of your humans is key. At the start of Season Five, most of the humans on our farm were once again skill-less and unproductive.

We had Alison that year, heavily tattooed and pierced, with blue hair, which Ella found fascinating. There was Andrea, the scientist, and Jason, who was very shy but also quietly competent. He would go on to work for 100km Foods. There was Melanie, who was strange and annoying, and Neil, who came later in the season and already knew how to work. And Katie came back for a second season as our "senior intern." Katie had originally signed up for a season on the farm on a whim, hoping to escape her corporate job and figure out what to do with her life. She ended up loving it so much that she decided she wanted to be a farmer. We were thrilled to have her back.

Katie helped a great deal with the training, but it once again took months to get the interns up to speed. I was feeling less and less enthusiastic about the time we had to spend managing feelings, accommodating needs and catering to whims. There's a common tendency to attribute character flaws to generational deficiencies; it's easy to say that our interns felt entitled and were soft and averse to hardship because they were millennials, but that's probably bullshit. I think twenty-somethings have been entitled, soft and averse to hardship for quite a few generations. Almost all of our interns were incredibly conscientious and hard-working. But they were all in their twenties. They had a lot of feelings. They had many whims. They did a lot of stupid things.

In Season Four, Todd inexplicably ate a bunch of red berries he found in a weed patch. We did some investigating and realized that they were deadly nightshade. It's always alarming to discover that one of your employees has eaten a plant that has the word "deadly" in its name, but poison control assured us that he would be fine if he'd eaten less than ten berries. He had, and he was. Also that season, Gillian was confronted by a feared intern delega-tion, this time to complain that Kirsten was keeping drugs in the communal fridge at the schoolhouse. It turned out that her family was involved in some kind of nature cult, and the drugs in the fridge were vials of peyote extract that they used in their rituals. We asked her to keep them at home. At the end of the season, at the height of the beet harvest, Susan suddenly left one day. She sent word that her boyfriend had broken up with her and she was far too upset to work anymore.

These kinds of things continued in Season Five. Alison informed us after two weeks of work in the spring that she had severe carpal tunnel syndrome in both wrists that prevented her from doing most manual labour. Gillian and I found it curious that this had not come up during the interview process, when we had asked Alison questions such as "How do you feel about doing hard physical labour all day?" We spent several weeks driv-ing her to physiotherapy appointments in Collingwood twice a week. (Alison, like almost all our interns, didn't have a driver's licence.) We eventually sent her home.

Later that season, Melanie approached me on a Friday morning, tears in her eyes. "My grandmother is dying," she told me earnestly. "I have to go home to Hamilton and sit vigil with her." I told her to take as much time as she needed. It was a harvest day in July and we were completely slammed, but we would make do without her, somehow. As soon as she left, the other interns started receiving Facebook updates. "I'm in town tonight, ready to party!" she posted. Melanie had left me her grandmother's number, so I called the next day to see how things were going. Grandma had evidently made a miraculous recovery, because she answered the phone and sounded much better than almost dead. She hadn't seen Melanie, she told me, and didn't know when she would be back. I wasn't sure what made me more angry, that Melanie had lied to us or that she had done such a bad job of it. When I finally got hold of her later that weekend, I told her not to bother coming back. The incident ended with a call from Melanie's mother, demanding to know why we had fired her daughter. "Melanie is in her mid-twenties," I pointed out. "She's an adult. I'll talk to her about why she was fired, but I won't talk to you." Now that I think of it, that's one character flaw that I can attribute to millennials; I never would have let my mother call my boss when I was that age.

BY THE MIDDLE OF OUR FIFTH SEASON of farming, we had rejected or come to question many of the major tenets of the small farm orthodoxy. We sold the large majority of our produce wholesale, rather than direct to consumers. We were increasingly specialized, growing fewer crops each year. Our farm kept getting bigger, and we were ever more dubious about the long-term viability of relying on interns for our farm labour. But at the same time, we were becoming more and more aware of the deep and pervasive problems with large-scale industrial agriculture. We were an organic island in a sea of conventional farms, and what we saw all around us was often shocking. If we had some reservations about the small farm orthodoxy, we found the industrial farm orthodoxy utterly terrifying.

For starters, very few seemed to be making a living at it. When we first decided to start farming, we met with our accountant, an old friend of my father's named Jim Boyko. Jim always looked haggard and world-weary, but he had a dry and acerbic sense of humour that I really enjoyed. At the end of our meeting he looked at us over his reading glasses. "You realize that no one makes money farming," he said slowly. I laughed. "Yeah, that's what people always say," I returned. I thought he was joking. "No," he said. "I'm serious. I work for a lot of farmers, and none of them makes money. Not one." Gillian and I just sat there uncomfortably.

A few years of living in a farming community confirmed Jim's observation. Virtually every conventional farmer we knew worked off-farm to make ends meet. Dan Needles, the playwright and essayist, who lives outside Nottawa, just north of us, sums this situation up succinctly. "Sustainable agriculture," he quotes his father-in-law as saying, "means a job driving a snowplow for the township in the winter and a wife teaching school."

Almost all the farmers in our community grow field crops—primarily corn, canola, soy and wheat—and are squeezed between high costs and fluctuating prices. The infrastructure and equipment needed to farm on an industrial scale is extremely expensive. Herbicides, pesticides and genetically modified seeds all cost a lot, and a big combine harvester can go for close to a million dollars. On the other side of the equation, commodity prices are set on the international market and change according to global supply and demand. A bumper crop of soybeans in Argentina, for example, can drive down prices all over the world and make the difference between profit and loss for a farmer down the road from us. Beef farmers in our community were all but wiped out soon after we started our farm when BSE, or mad cow disease, was detected in Canada and the U.S. border was suddenly closed to Canadian beef imports.

The end result is farmers who have millions of dollars invested in their operations yet still have to drive a truck or plow weekenders' driveways to pay the bills. As Gillian and I learned more about the economics of agriculture, we began to see evidence of this paradox all around us: a neighbour with two hundred thousand dollars' worth of tractors parked in the yard who picked up work cleaning houses. Another who spent more than a million on a new hog barn yet lived in a hovel. Farm after farm where the buildings that housed the machinery and livestock were nicer than those that housed the farmers and their families.

The economic problems with industrial agriculture were plain for us to see, but the environmental problems were even worse. The large majority of conventional farmers in our community are conscientious, responsible stewards of the land. They might farm with large machinery and a lot of chemicals, but at least they make an effort to minimize their impact. There are a few, however, who might best be described as environmental vandals. I once walked a property where the farmer kept his beef cattle in a concrete yard in the winter and sluiced the runoff from the manure directly into the Batteaux Creek, an otherwise pristine trout stream. I've seen farmers burn or bury all manner of noxious garbage on their property, apparently unconcerned about the danger of contaminating their own water, air and soil. Many times I've witnessed farmers spraying their fields in high winds, in contravention of both the law and common sense.

But the scariest thing we saw was the pesticides. We would get bulletins from farm organizations listing dozens and dozens of different chemicals that had been approved for vegetable crops. Some of them were known carcinogens that had been banned years ago, but farmers could apply for temporary use permits to handle an outbreak of a specific insect or disease. The list of pesticides approved for use on spinach and lettuce is pages long, yet these vegetables are harvested only three or four weeks after planting, so the interval between spraying and eating can't be very long. I saw the millions of flea beetles crawling over the outside of our row cover and tried to imagine how drenched in chemicals our arugula would have to be in order to keep the bugs off.

One day in the spring of Season Five, Gillian and I were taking a walk on our sideroad after dinner, as we often do. The big field on the other side of the road had been planted that day—we had seen the enormous tractor towing an equally enormous seeder. We noticed some seed that had spilled on the road. They were soybeans, but they were an unnatural colour, almost fluorescent red. I picked some up and looked at them more closely, and I could see that the red colour was some sort of powdered coating. I quickly dropped the seed. "This is the stuff we've been hearing about," Gillian said.

The previous winter Gillian and I had attended the Ontario Fruit and Vegetable Growers conference, the first time we had signed up for a conference that didn't focus exclusively on organic production. The trade show was dominated by chemical and equipment manufacturers, and there was a boozy reception sponsored by Bayer Crop Science, one of the world's biggest pesticide companies. One word kept appearing on all the posters and glossy brochures at the conference: "systemic." I asked a company rep at one of the booths what it meant, out of curiosity. He looked at me like I was stupid, a look I had grown accustomed to by then. "You know, systemic," he said. "It means it works through the whole plant. The whole plant is toxic to insects."

But it wasn't just insects. There were booths flogging systemic fungicides as well. The literature we picked up bragged about the wide range of fruits and vegetables that could be sprayed, and the length of time that the chemicals remained active after spraying. An insecticide for apples came with the assurance that the fruit would remain toxic to crawling insects for several months after application. One fungicide for fruit was designed to prevent spoilage *after* harvest, right up to the point of sale, while another was injected directly into the trunk of the tree to "protect" the fruit, leaves and roots.

Creating a pesticide that works by making the whole plant toxic, including the part we eat, struck both Gillian and me as a profoundly bad idea. But industrial agriculture seems to double down on bad ideas. When we got home from the conference and did some research, we discovered that sales of systemics are growing faster than sales of any other class of pesticide. The most popular family of systemics, and the one most people are likely to have heard of, is neonicotinoids, or neonics, for short. The soybean seeds we picked up on the road that spring had almost certainly been treated with neonics. One form of neonic—imidacloprid—is now the bestselling pesticide in the world. The reason neonics are so popular is their incredible persistence—applied as a powder to a seed, they pervade all the tissues of the plant as it grows and make every part of it toxic to insects for the entire life of the plant. This remarkable persistence is also the reason neonics are so destructive. They can remain active for several years, creating toxicity not just in the plant they were originally applied to, but also in wild plants growing nearby and in subsequent crops grown in the same field. Neonics are toxic to virtually all invertebrates, so they kill indiscriminately, and they are water-soluble, so they can travel freely through the local environment. I thought about the bees and butterflies and innumerable other insects that inhabited the ditches and wild buffers around our farm, and of the bright red seeds scattered across the road.

"If only people knew," Gillian said as we walked home that day, vainly kicking seeds to the far side of the road. It was a thought that came to us often back then as our eyes were opened to the realities of conventional agriculture. Later that season a woman we had never met before drove into our lane and asked if she could buy some potatoes. She seemed a little nervous as Gillian loaded her car. "I can't let anyone see me," the woman said in a conspiratorial voice. "My family has a big potato operation just south of here." She explained that her relatives often sprayed their fields fifteen times in a single season, including a dose of herbicide to kill the potato plants before harvest, so the vines wouldn't clog up the enormous harvesting machines, and a shot of another chemical after harvest, to prevent sprouting in storage. "I know," she said before she drove off. "I've seen what they do, and I won't eat it."

OUR GROWING KNOWLEDGE OF the environmental and economic problems created by industrial agriculture intensified our determination to find a better model, but these same problems made it harder to farm sustainably. Farmers who subsidized their money-losing operations with off-farm work helped keep food prices low, which undercut our ability to make a profit. And the large-scale environmental impact of industrial farming was degrading the quality of the water, the abundance of pollinators, even the stability of the global climate—in short, all the natural systems necessary to grow food.

The desire to do something about the looming threat of climate change had been one of our original motivations for starting the farm. We had become convinced that changing agriculture is key to altering our impact on the climate. By Season Five, however, it was clear that the agriculture–climate interface works both ways; agriculture contributes greatly to climate change, but climate change, we began to see, was also changing agriculture.

The first issue was water. We had bought a drip irrigation system early in Season Two because that was Eliott Coleman's recommendation. Drip irrigation uses far less water than overhead sprinklers, so we could hook it up to our house well without worrying about running it dry. And since drip lines are just light plastic hoses with drippers built in every twelve inches, they are relatively easy to move around on a small farm. But Seasons One and Two were wet and rainy, so we never needed to water anything. Season Three was a lot hotter and drier, and in Season Four we experienced our first real drought. All the computer models predict changes in the amount and timing of precipitation as the climate warms, with temperate areas experiencing more frequent dry spells in the summer. In 2010, we had sixty days in June and July with only one significant rain. Our fields were bone dry, and it was hot and sunny almost every day. Drip irrigation is designed primarily for the slow and steady watering of long-season crops. It worked great for our beets and carrots, but in drought conditions it was useless for salad. We were planting so many rows of salad by then that we had to lay down dozens of runs of drip-line to water a single week's planting, and the light hoses blew all over the place when they were empty between waterings.

With salad greens, the crucial time for watering is immediately after planting, so during the drought of Season Four we jury-rigged a contraption that could water a planting of greens in one go. I used flexible plumbing pipe and four ordinary garden sprinklers to build a rig that could water a 200-by-30-foot swath of garden in a couple of hours. It leaked a lot and it was a huge pain in the ass to move, but it worked, more or less. Katie dubbed it the Germinator.

We struggled with the Germinator for a few years, but as time went by, the pattern of hot, dry summer weather intensified. In Season Six we had to irrigate in April for the first time, and we had to irrigate just about every planting of salad for the whole season. We still got lots of rain, but it came in short, violent thunderstorms that would dump several inches of water on the farm in the space of an hour or two. Then we would go two or three weeks with no rain at all. We were running the Germinator so much that I worried we would eventually exhaust our little house well, so we finally invested in a proper irrigation system.

Rubin came over with his big excavator and dug a huge pit in a wet part of the field on the far side of the barn. It immediately filled up with groundwater. Then I rented a ditching machine and dug trenches to all our fields. Dan came up from Dunedin and helped me glue together over two thousand feet of pipe, which we laid in the trenches and buried. We hooked up a small gas pump by the pond, and we were in business. The underground pipes had a riser in each field where we could hook on big green hoses to get the water where we needed it. We would lay out a run of light aluminum pipes with big sprinkler heads attached, down the middle of newly planted sections of the garden. Now we can hook up the pipe, fire up the pump and water a 200-by-160-foot swath of garden in about an hour. When I meet aspiring farmers these days, I tell them that a good irrigation system is essential, right from the start. The days of planting your vegetables and then hoping for rain are over.

Changing weather patterns have also had an impact on what crops we grow. Tomatoes were one of the things that got us interested in farming in the first place. Nothing illustrates the vast difference between the generic tastelessness of supermarket produce and the delicious diversity of homegrown vegetables better than the tomato. Tomatoes also like hot weather, so we thought a warming climate might actually increase our tomato-growing success. We made a large investment in Mennonite fencing to trellis our heirloom tomatoes back in Season Two, and later we started planting up to half an acre of multicoloured field tomatoes every season. Our healthy organic soil ensured vigorous, healthy tomato plants and heavy yields, with no sign of pests or disease. Then came the blight.

Late blight is a fungal disease that overwinters in the warm southern United States and is carried northward each summer on the winds. In the past, it usually arrived in northeastern North America in the cool, wet weather of the late fall. Late blight can turn a healthy tomato plant into a withered brown mess in just a few days, and it causes ripe fruit to turn blotchy and eventually rot. There is very little that conventional growers can do about late blight, and organic growers are essentially helpless.

In the early 2000s, organic growers all over the northeast noticed that late blight was arriving earlier in the season as winters grew less severe. Then in June and July of 2009 the blight rampaged through tomato gardens, big and small, all over New York, New England, Ontario and Quebec, spread not only by warming winds but also by contaminated seedlings that had been grown in the southern United States and trucked to big-box garden centres all over the continent. Our tomato garden was very small at that time, and we were mercifully spared. In 2011 our field tomatoes were hit in mid-September, but we had already sold most of the fruit by that time. Then, in late August 2012, just as our field tomatoes were hitting their peak of production, we noticed a few brown leaves on some of the plants and the odd spot on the ripening fruit. Within two days the whole field was withering. In less than a week, everything was dead. We estimated that we lost four thousand pounds of tomatoes to the blight—almost our entire harvest—and we gave up on growing tomatoes after that season. The combined forces of climate change and corporate consolidation had driven us out of the tomato business.

WHEN IT CAME TO THE MYRIAD PROBLEMS with our chemical-industrial food system, Gillian and I had a front-row seat, but we weren't the only ones who noticed that something was wrong. It was clear by 2011 that we were in the midst of a sea change in the way we thought about food. When we had started farming, the idea that people should care about how food is produced was new and revolutionary. Now, five years later, that idea was firmly in the mainstream. Sales of organic foods were increasing by 15 percent a year, the fastest-growing food category, and big retailers like Costco and Walmart were getting into organics in a big way. We saw the change up close and personal in the farmers' market. Our clientele had at first been made up almost entirely of weekenders and retirees, people with lots of money. Many of them were committed organic consumers, but many also seemed to shop in the market because that was the thing to do.

But as the years passed, we saw more and more ordinary people buying from us, folks who worked at the Honda plant down in Alliston, working-class families with young kids, old ladies who had lived in Creemore all their lives. They all had a story to tell of what brought them to our stand, and it almost always had to do with a health problem that had touched their lives. A husband with diabetes. A child with colitis. And story after story of cancer. I can't tell you how many times customers would come to our stand, pick out their vegetables and then lean in as they were paying and tell us about someone they loved who had cancer. "My mother is too sick to come to the market," one middle-aged woman confided to us one day. "But she loves your eggs. She says they taste like eggs did when she was a little girl."

My mom had been diagnosed with lymphoma several years earlier, soon after we started farming. Lymphoma is a form of cancer closely associated with exposure to herbicides; it's most common among forestry and golf-course workers, and farmers. My mom confronted her cancer in the same way that many of our market customers did: she radically changed her lifestyle, did more exercise and sought out organic foods. Keeping her stocked with vegetables, eggs and meat from the farm made me feel a little less helpless, in the same way that making our food available to our friends and community members in the market made Gillian and I feel that what we were doing was important. We felt like we had a tiny bit of agency in the face of the tidal wave of diet-related illness that seemed to be engulfing our whole society.

Late that fall, at the end of Season Five, I made the drive to Hanover to see Gerald Poechman, to pick up some feed for the hens we were keeping over the winter. After we had bagged and loaded the feed, we stood around the back of my pickup and talked, like we always did. I told Gerald about the changes we were seeing in the market and among our restaurant clients, about how people in all walks of life were starting to make the connection between the food they ate, their own health and the health of the planet. I told him that I thought we were reaching a tipping point, where real change in our food system was possible. And I told him that all these changes had conspired to create the conditions where our little farm, using mostly hand tools and a workforce of commit-ted, idealistic young people, could actually turn a profit. Gerald smiled. "I always knew you could do it," he said.

Then Gerald told me about the changes he was seeing in his community. He told me about the conventional egg farmer down the road whose laying barn was so highly automated that he could offer his lone worker only twenty hours of work a week. He had recently doubled the size of his operation, to a hundred thousand hens, so he could bring his hired man on full-time. The worker spent most of his day walking the rows of cages, taking out dead birds. Everything else was done by machine. Gerald also told me that many of his conventional neighbours were cutting down the trees and ripping out the fencerows between their fields to accommodate ever larger machinery. They were using enormous sprayers with sixty-foot booms that could cover an acre of crops with pesticide in a matter of seconds. The landscape was being transformed into massive, unbroken fields, hundreds of acres of perfect, sterile monoculture. Globalized commodity markets and razor-thin margins meant fewer and fewer farmers and bigger and bigger farms. "You're part of a tiny little segment of agriculture that is moving in the right direction," Gerald told me. "The rest is hurtling in the opposite direction, as fast as it can."

As I drove home through the rolling hills along Grey Road 4, I turned over Gerald's words in my head. We had started our farm with a mission to prove that a small farm could be both profitable and sustainable. We had finally succeeded on the profitability side, sort of. Our Season Five sales had grown only modestly over Season Four. We had made enough to cover all our farm expenses and to live for a year, but only because our living expenses were so low. We didn't spend much money because we were working pretty much all the time. But on the sustainability side, I suddenly realized, we had failed miserably. I truly believed that our farm was ecologically sustainable—our soil was improving each year, we had planted thousands of trees on parts of our farm we weren't using to soak up carbon dioxide and provide wildlife habitat, our solar panels made us self-sufficient in electricity and we didn't burn a lot of diesel—but we had been thinking of sustainability far too narrowly. For our farm to be truly sustainable, it had to sustain Gillian and me physically, intellectually and emotionally; it had to provide both a livelihood and a life. Our farm had to protect and enhance not only our air, water, soil and wildlife, but also our family. For our farm to be sustainable, it had to be sustainable for us. And at that point it wasn't.

There was no way we were going to build a new model of agriculture that required farmers to destroy their bodies, ruin their marriages, forsake their intellectual interests and cut themselves off from their friends and family in order to turn a meagre profit. No one was going to leave the city and follow in our footsteps if they saw what we had gone through for the past five years, and we certainly weren't going to convince any conventional farmers that we had figured out a better way. "The rest of agriculture is hurtling in the opposite direction," Gerald had said. If we were going to make real change, we needed a viable alternative. And at that moment, despite five years of grinding toil, our farm wasn't it.

CHAPTER 10

LOS MUCHACHOS

AT THE END OF SEASON FIVE, Gillian and I did what we always do at that time of year. We collected all the data from the previous season in big spreadsheets—sales, expenses, yields, field observations—then sat down for a full-day planning meeting. We always hold our year-end meeting in November, and it always seems to be cold and rainy outside. That year was no exception. We stoked up the woodstove, gathered our spreadsheets and sat down to hash things out. It was immediately obvious that we needed to make big changes.

When I meet people who aren't familiar with our farm, the first question they ask is always the same: "How much land do you have?" In just about everyone's mind, farm scale is measured in acres. It took a long time, but by the end of Season Five, Gillian and I had figured out that this was the wrong question to ask. On a human-powered farm like ours, the limiting factor will always be labour. We have lots of unused land on our hundred-acre farm that we could expand onto, and we could always rent more. None of our immediate neighbours are farmers—they rent their land to conventional farmers, some of whom live far away, which is the norm in North America. Access to land isn't what determines how much we can produce. On a human-powered farm, the number of workers is the best measure of scale. The right question isn't "How much land do you have?" It's "How many workers?"

When we looked at our numbers that year, it was clear that we had reached the limit of what we could do with interns. We had brought on as many as we could manage, gone to elaborate lengths to make them happy, and invested in equipment to make their work as efficient as possible. But our sales growth had stalled. We were profitable on paper, but we were only making enough to get by. We couldn't bring on more interns and I didn't see any way to make the ones we had more productive. In terms of growth, we were at a dead end.

At the same time, our farm was becoming less and less attractive to interns. We had become more specialized over the years as we stopped growing stupid vegetables, we were becoming less dependent on the farmers' market as our wholesale business grew, and we were getting ever bigger. Whatever we thought of the small farm orthodoxy, most prospective interns were wholly enamoured with it. Their ideal was usually a very small, very diverse farm that ran a CSA. But we were not following that model.

During Season Five, we had collected more detailed data on our sales and labour inputs than we had in the past, and the story that data told was stark. We made twice the return on labour selling wholesale as we did in the farmers' market. The market, we realized, was a giant labour suck—we spent enormous amounts of time bunching and prepping our vegetables, then spent all day selling. Sales in the market were so sporadic that on a rainy day we might end up throwing out half the vegetables that we had spent so much time preparing, and on an unexpectedly busy day we might sell out two hours before closing and lose out on a lot of potential sales. Our wholesale orders were always sold before we harvested, required far less labour to produce and then went straight into the 100km truck. From a financial standpoint, there was no comparison. We were also worn out after spending every summer Saturday in the market for five years. We wanted to spend weekends with our kids, maybe even go to a friend's cottage once in a while. It was painful, but we decided that our time in the farmers' market was over.

The data also told us that almost all our profit came from just four crops—beets, potatoes, cucumbers and cut salads. These were the things that grew well on our farm, that we enjoyed growing and that sold well. If we wanted to take our specialization to its logical conclusion, those four crops were it. Gillian and I agreed to get rid of everything else, but we had a knock-down, drag-out battle over the carrots. Our multicoloured carrots were beautiful and much loved by our customers, but we didn't sell very many bulk carrots to restaurants; most of the demand was for bunched, and bunching takes a lot of time. Gillian was very reluctant to give them up, but we did an analysis at the end of the season and determined that our return on labour for bulk salad was *five times* what it was for bunched carrots. In the end, carrots got the heave-ho.

So where did that leave us? We couldn't grow our business with interns, and they wouldn't want to work on a specialized, wholesale-focused farm, anyway. We wanted to increase the scale of our human-powered farm but we couldn't get more workers, so we needed different workers. We needed better workers.

The first time I'd met Morris Gervais of Barrie Hill Farms, he'd asked me a whole bunch of questions about our farm, then looked me in the eye and said this: "When you're ready to get serious about farming, get Mexicans." Every time I saw him after that he would ask the same question: "Are you ready to get serious?"

He thought we were crazy to run a farm with a bunch of unpaid twenty-somethings as workers. After five years of farming and three years of interns, we finally had to admit that he was right. Gillian called him up immediately after our meeting. "Morris," she said, "we're ready."

WHEN GILLIAN CALLED MORRIS and told him we were ready to hire Mexican workers, he sounded like he would jump for joy. Morris employed about forty migrant workers, all from Mexico, and he struggled to find the superlatives to describe what wondrous workers they were. His workers obviously liked him, too; some had been with him for more than a decade. He had taken a Spanish language class one winter and spoke fairly fluently by the time we met him. He gave a prize each year to the worker who taught him the best new Mexican profanity; the winning phrase that year had been *culo bajito*, used as a nickname for a short person, which literally translates to something like "low-altitude asshole." Morris told us that each Mexican worker would be able to do the work of two interns, so we decided to hire four.

We resisted hiring Mexican workers for a long time because, like most people, our image of migrant farm workers came straight out of *The Grapes of Wrath*. We assumed that Mexican workers must get a raw deal, that the relationship would naturally be one of exploitation and abuse. This is often the case in the United States, where around 70 percent of farm workers are undocumented migrants. It's common for workers to follow the harvest from state to state, living in their cars or in illegal shanty settlements, making minimum wage or less, with no benefits and few legal protections. But the more we saw of Morris's operation, and the more research we did of our own, the more we realized that the situation in Canada is different.

In Canada we have the Seasonal Agricultural Worker Program, run by the federal government, which allows farmers to bring in seasonal workers from countries that have signed on to the program—mostly Mexico and Jamaica, but there are a bunch of others. Big parts of our agriculture system are entirely dependent on these migrant workers, including almost all fruit and vegetable production. Ontario alone receives about eighteen thousand workers under the program every year. The whole thing is very tightly regulated, and while they're here, workers receive the same benefits and protections that Canadian workers receive—full health coverage, guaranteed wages and federal benefits. Some workers accumulate enough months of employment to receive a Canadian pension when they retire. Several of our workers have claimed Employment Insurance benefits for paternity leave when they had a baby born back home in Mexico. The system is far from perfect, but we thought we could give our workers a fair deal.

Before we could join the program, our worker housing had to be inspected by a public health nurse to ensure that it met the standards. We also had to prove that we couldn't find Canadians to fill our positions. We advertised on a green-jobs website and sent word out to the network of former CRAFT interns, but without a CSA or a farmers' market, none were interested, even though we were offering a salary. The applicants we got through the government job bank seemed interested only in demonstrating that they were looking for work so they could maintain their unemployment benefits; none of them returned our calls. Many farmers see advertising for local workers as a hassle, an administrative hurdle they are obliged to jump over, so they put in minimal effort. We really tried, but to no avail. I suppose we could have found some Canadian workers if we were paying significantly more than minimum wage, but that would have forced us to make our vegetables even more expensive than they already were. To pay enough to attract local workers, while competing with an industrial food system bent on keeping prices low, was a financial impossibility.

All of the regulations and requirements for the workers added to our expenses. We were required to pay everyone minimum wage, which in 2012 was about $10.50 an hour. But we also had to provide free housing, pay for half their airfare and work visas, provide transportation on their days off, pay for supplemental health insurance and cover their payroll benefits. All this raised the cost of employing offshore workers to about $17 an hour. Gillian calculated that we would have to increase our sales by about $75,000 just to cover our increased labour costs. That seemed like a very tall order.

I GOT UP VERY EARLY ONE MORNING in April 2012 and drove the hour and a half to the airport in Toronto. I parked, then followed the directions I had been given, down the escalators from the busy arrival and departure areas to the mostly deserted basement of the terminal. I followed a corridor to a large room where I found a group of about a hundred men sprawled on chairs amid piles of suitcases and boxes. A man with a clipboard and a walkie-talkie waved me over. "What farm?" he asked briskly. "The New Farm," I told him. "Your guys are over there." He pointed to a knot of four men standing in a group off to the side. I walked over and we shook hands. They all appeared to be in their early to mid-thirties. They had been travelling for almost twenty-four hours, having left their home village in the central Mexican highlands by bus the previous morning, then flown overnight from Mexico City, so they were bleary-eyed and quiet. We mumbled some greetings, gathered their bags and headed for the car. This was my first meeting with the Perez-Victoriano brothers—Juan Carlos, Chuen and Luis—and their friend Rene.

The five of us crammed into the pickup for the ride home from the airport. We stopped at a greasy spoon along the way and I bought everyone breakfast, which we ate in silence. The three Perez-Victoriano brothers had worked on farms in Ontario for a number of years, and they had all worked for Morris at one time or another. Morris had told us that they spoke a little English, but I didn't see any evidence of that on the ride home. I knew a grand total of three words in Spanish. It dawned on me during our silent ride that we were taking a huge financial gamble on workers I couldn't communicate with. I started to wonder if we had made a big mistake.

I picked everyone up at the schoolhouse the next day and drove them the half mile to the farm. The fifth and final member of the crew had arrived the previous afternoon. Gillian and I had anticipated the language problem, so on Morris's advice we had signed on with Frontier College, a national literacy organization. Frontier recruited, trained and placed young people to serve as labourer-teachers, something they had been doing for more than a century. The program started off with idealistic young people living and working in lumber and mining camps in the Canadian north, teaching illiterate workers to read and write. The focus had shifted to migrant farm workers by the time we got involved, but the method was still the same. We ended up with a guy named Liam who had just graduated from university and was in his early twenties. Liam would live and work with the guys and we would pay him, just like everyone else. In the evenings, he would teach them English.

It turns out that Mexicans are big on nicknames. That first morning, just about everyone got one. The Mexican nicknaming process is free of even the tiniest hint of political correctness. Rene was a little heavier than the others, so he was Gordo, which means "fatty." Liam was Guero, which is the Mexican equivalent of "honky." Every labourer-teacher on every farm in Ontario is invariably known as Guero. The guys would sometimes refer to Liam as "our *guero*" to distinguish him from *gueros* on other farms. Luis was Wichos. I have no idea what that means. The crew referred to themselves collectively as *los muchachos*, meaning "the boys" or "the guys." I was *patrón* and Gillian was *patrona*. That made me uncomfortable. In books I had read, the *patrón* was always the hacienda owner or the plantation master. The *patrón* was the bad guy.

It was mid-April and very wet that day, so there wasn't much we could do in the fields. I took the crew into one of the hoop houses and showed them how to kill all the weeds and grass along the sides with a hoe, in the areas I couldn't reach with the tiller. It would have taken the interns several hours to do that job, so I left them and went off to do some work on the tractor. Liam found me in the barn less than an hour later. "We're finished," he said. "What's next?" I told him to do the same thing in the other two hoop houses. "No," he said. "We've already done them all." I thought he must have misunderstood me, so I walked over to the hoop houses to check. All three had been weeded perfectly. That would have been a full-day job for the interns, and I would have had to work with them the whole time.

I spent the rest of that day scrambling to find enough work for everyone; I was relieved when five o'clock rolled around so I could send them all home. Liam let the others know that the day was over, but they didn't look at all happy. It took a few minutes of hand gestures and broken Spanglish for me to figure out what was upsetting them, but the gist of it was this: "When we sit at home, we think about our families and we get sad, so we would rather be working. We are here to work."

I called Morris that night to ask him what I should do. He laughed. "I told you!" he said. "They're not going to be happy with less than sixty hours of work a week. You better start planting more than you think you can harvest, or you'll run out of work." Gillian and I quickly decided to double our salad plantings. Keeping interns happy often meant limiting the number of hours we worked, but now there was a new imperative on the farm: keep everyone working as much as possible.

The next few weeks were an exercise in creative work-making. The crew picked rocks, mucked out the chicken coops and prepared all the hoop houses for planting cucumbers. I trained Rene to make soil blocks, then he taught everyone else. Once we ran out of actual farm work, we started landscaping. The guys added cedar rails to the fences around the barnyard, which made it look much nicer, and built rock borders around the flower gardens. When I got desperate and asked them to put a stone border around the fire pit, they found hammers and chisels in the barn and shaped all the rocks, so they could create a perfect circle. We had gone from workers who had never used a hammer to workers who were experienced stonemasons. Growing up in a village in Mexico and working on farms for years had given these guys a range of practical skills that very few Canadians could match. They knew how to work, and then some.

THE EXCITEMENT GILLIAN AND I felt about our new employees was tempered by the knowledge of how much we were spending on them. Our last root vegetables had been sold the previous December, and we would have no income at all from the farm until well into June. We usually started selling salad by the third week in May, but we didn't actually get paid until thirty days after delivery. That meant almost six months of the year with no money coming in. Meanwhile, our employees were costing us over eight hundred dollars a day in wages and benefits. Before we harvested a single leaf of lettuce, we had already paid our five employees more than we had paid our eight interns the entire previous season. Watching that much money leave our account every day, with nothing coming in, caused a great deal of stress.

That stress manifested itself in the usual way—I acted like an asshole. After three years of treating interns with kid gloves, I somehow felt that having paid employees gave me the right to lose my temper and raise my voice when I wasn't happy with their performance. I didn't have to worry about anyone's feelings anymore, I told myself. These were grown men, so if they fucked up, I let them know it. The fact that we didn't speak a common language made fuck-ups inevitable. Our communication problems were compounded by the speed at which the guys worked. When they did something wrong, they did it wrong fast, so a lot of it got done wrong before Gillian or I realized. I also felt that the guys were deliberately slowing down when we didn't give them as many hours as they wanted. I got angry when they didn't work fast enough, but they were working too fast for the amount of work we had to do. It was ridiculous. Gillian tried to keep me under control, but I spent a lot of my time being angry, despite her best efforts. I'm not proud of the way I acted back then.

My outbursts were met with downcast eyes and mumbled apologies by everyone but Luis. Luis is the youngest brother, and he carries himself with a kind of defiant pride. When I started yelling, he would stare intently into my eyes with a look that bordered on malice. He never backed down or apologized. Sometimes he would goad me. "You lookin' very ugly when you hangry," he would say in broken English, with no hint of humour. That made me even madder.

One day, about a month after the guys arrived, Luis asked if I would drive him to the farm he had worked at the previous season, so he could pick up some things he had left there. The place was less than an hour away, near Barrie, so it wasn't a big deal. Luis and his brothers were getting more comfortable on our farm and more willing to try out their rudimentary English; their evening English lessons were also starting to pay dividends. Poor Liam would be exhausted from the day's work, but the guys always went straight for their notebooks when they got home to the schoolhouse. "*Vámonos*, Guero!" they would say. Let's go, Honky! They were eager to learn. I was also picking up a few words of Spanish, so Luis and I were able to understand each other as we talked on the way to his old farm. He told me about his family back in Mexico—his wife, his beloved daughter, who was about the same age as Ella, and his son, Luis Junior, who had some sort of disability. "No walk, no talk," was all Luis said. They lived in the village that Luis and his brothers had been born in, on the slopes of a beautiful volcano called Popocatépetl. He laughed as I tried again and again to pronounce the name properly and failed.

There was no one around when we arrived at the farm; the farmer lived on another property down the road. The old farmhouse where Luis had lived was used only for housing workers. We went inside and I caught my breath. The place stank of mould, like a wet basement. There were holes in the plaster and water stains on the walls. It was incredibly dirty. One room on the ground floor was completely stuffed with piles of old furniture and broken junk. We walked up the dark, narrow staircase and into Luis's old room. There was a steel cot with a stained mattress and no other furniture. The old beige wallpaper was peeling off the walls in several places. I couldn't figure out how this place had passed inspection with the public health department. Luis's stuff was still there, a garbage bag with some rubber boots and winter clothes. We grabbed it and hurried out.

Before we left, Luis walked me back to where he had worked. Behind the house there was a large pit, about a hundred yards across and ten deep, that looked like an old quarry. In the bottom there were piles of rocks, some rusty equipment and a battered old shipping container. Luis's job had been literally to break rocks, using sledgehammers and heavy saws to reduce large boulders to flagstones and pavers. The farmer ran a landscape supply business on the side. Luis had been instructed to hide in the shipping container whenever anyone visited the job site. It was illegal to use migrant agricultural workers for non-farm work, and his old boss didn't want to get caught.

I was outraged. "We have to report this guy!" I said as we drove away. But Luis was philosophical. *"Muchas horas,"* he said—lots of hours—and the farmer always paid cash. For Luis, the real hardship was being away from his family for so long. He was here to make money and support his loved ones back home. If that meant living in a shithole and breaking rocks twelve hours a day, then so be it.

A week or so later, we had a visitor from the Mexican consulate in Toronto. One of the consular officials visited each new farm in the program to make sure everyone was fulfilling their obligations. The guy who came that day was tall and well-dressed and spoke impeccable English. He first met with the guys, then met with Gillian and me, then brought us all together—shuttle diplomacy on the smallest possible scale. He spoke to us in the manner of a schoolmaster addressing a bunch of kids who had been fighting on the playground. He said everything first in Spanish, then in English. "You must work quickly, even if you aren't getting as many hours as you want," he said to the guys. Then he turned to me. "And you must stop yelling." He told us that we were all adults, that we all deserved to be respected, and that we had contractual obligations to each other that required us to work together. I felt a strange mix of shame and embarrassment. I had acted like a dick and was being called out for it by an actual diplomat. I caught Luis's eye, and he winked at me. He was clearly enjoying watching me squirm.

WE CAME TO AN UNDERSTANDING that summer that still holds on our farm: the guys would work diligently and quickly and we would provide as many hours of work as they could handle. It turned out that they could handle a great many hours. We planted more and more, but the guys still managed to stay on top of everything. For the first time in the history of our farm, we were pretty much always caught up. This made our whole operation much more efficient, because we were never expending the extra labour required to catch up on a task delayed—there was no hacking away at overgrown weeds or pruning three-foot suckers off the cucumber vines. Everything happened when it was supposed to. We even found time to cut the grass and tend to the flower gardens. The place looked more professional, and it was.

My job was to make sure the guys had everything they needed to keep working at peak efficiency. I fixed things when they broke. I ran to Hamilton Brothers for supplies. I did most of the tractor work, and Gillian or I did almost all the planting. I spent hours walking the EarthWay back and forth across the gardens, planting about three miles of row every week. Gillian planted all the beets and did lots of other field work, but her specialty was selling. Our Mexican crew was transforming the production side of the business, so we had to figure out a way to transform the marketing side to match.

With our higher level of production, we could now approach more, and higher-volume, customers. The produce distribution business is still fragmented and localized in Canada—just about every medium-sized town in the country has an independent, family-owned produce distributor; in Ontario, a lot of them are owned by Italian families. Gillian cold-called Sanfilippo's, a small distributor based in Collingwood that serves a wide swath of central Ontario. They followed the standard business model—bring stuff up from the food terminal in Toronto every morning, then deliver it to restaurants. They jumped at the chance to work with us. Their clients had been asking for local and organic, but Sanfilippo's didn't know where to find it. Domenic Sanfilippo, who seemed to single-handedly run the entire company, is a warm and generous guy; he started placing large orders right away. The local food movement has spent a lot of time and effort trying to dream up new ways to get food from local farms to restaurants and grocery stores—food hubs, cooperatives, Uber-like delivery services—but we found it quicker and easier to hijack the existing produce distribution system. Sanfilippo's soon became our second-biggest client, after 100km.

Gillian also made contact with Mama Earth Organics, a large home-delivery company in Toronto. At first they were skeptical about our ability to deliver the kind of volume they required, but Gillian assured them that we could meet their needs. They had about twelve hundred clients back then, so their first order was for twelve hundred cucumbers, our single biggest cuke order of all time. A few weeks later they took three thousand pounds of potatoes. We had been used to selling spuds in ten-pound bags. We never could have filled such a big order with interns, but the guys buckled down and harvested potatoes for fourteen hours straight. It was early August, and we finished up at 9 P.M., as the sun was setting. Luis smiled as we all walked out of the field. *"Muchas horas!"* he said approvingly.

By August it seemed like we spent every waking hour harvesting. 100km Foods was now picking up on the farm twice a week, on Mondays and Wednesdays. Sanfilippo's was picking up on Tuesdays and Thursdays. Mama Earth usually picked up on Sunday. Our orders came in early in the morning on the day of pickup; we had honed our just-in-time harvesting process so that everyone knew exactly what they had to do. We had also standardized our processes to minimize language-related fuck-ups. Gillian and I would transfer orders to a large whiteboard in the wash shed and write up adhesive tickets for each bin of produce that the guys needed to pack. These were pegged on an order board above the packing table, just like orders on the line in a restaurant. The harvesters would keep track on the whiteboard of what came into the wash shed, and the washers would pack the bins, label them and load them into the walk-in cooler. Gillian would write up the invoices and I would double-check them against what we had in the cooler. We were usually scrambling to wash the last of the order as the truck arrived. Then Gillian and I would load everything into the truck. Our cooler would fill up all day and empty each afternoon. Then we would get up the next day and do it all over again. We found time whenever we could for everything else that needed to be done—planting, covering, weeding, pruning. We started getting orders that added up to four or five thousand dollars a *day*. The money started pouring in.

THE GUYS ALL WENT HOME at the end of October. By the end of the season they had started referring to the farm as "our farm," and they all vowed to come back next season. They had fully embraced our commitment to producing the highest-quality vegetables, and repeated the mantra in Spanish over and over, every day: *primera chulada*—top quality. Juan Carlos often admonished me when I was tempted to cut corners in order to speed things up. "No, *patrón*," he would say in the hybrid Spanglish that we had developed. "*No es* good quality."

Employing migrant workers is a very serious departure from the small farm orthodoxy, and I could tell that some CRAFT farmers disapproved when we told them we would no longer be taking on interns. Gillian and I still had misgivings after that first season, and we were still bothered by nagging questions: Did bringing workers from thousands of miles away fit into a sustainable model of agriculture? Was a guest-worker system inherently exploitative, no matter how well we treated our crew? But the same questions could be asked of an agricultural system based on the unpaid labour of interns, and we had consistently failed to attract paid local workers, despite our best efforts. In the end, it was *los muchachos* themselves who settled the argument. They were skilled and competent workers and beautiful human beings who badly wanted to return. The pay and the working conditions were so much better on our farm than anything they could find in Mexico, it made the sacrifice of being away from their families worth it. We weren't sure if we were allowed to do it under the rules of the program, but we gave them all a big bonus when they left.

When we added up the numbers that fall, we found that our increased sales covered the extra labour costs and then some. We had made enough to cover all our expenses, to live for the year and to pay down a significant portion of the debt we had taken on to buy the schoolhouse and the solar panels. Finally, after six seasons, we weren't losing money or just getting by. We were honestly and truly profitable.

But the real difference came the next year. The labourer-teacher program was cancelled, so we couldn't get another *guero*, and Rene didn't come back, but Juan Carlos, Chuen and Luis did. Starting the season with an experienced crew was a revelation. The three brothers trained the two new Mexican workers we brought on, and started off right where they had finished the previous year. This time, we were ready for them, and we vowed to keep them busy all season, from start to finish. In the spring we put up two more hoop houses that were twice as big as the big ones we already had, and we launched ourselves into the retail market. We had sold cucumbers, beets and carrots to a couple of independent grocery stores in Toronto in previous years, but now we set ourselves up to bag salad for retail and actively sought out new markets. We hooked up with an all-organic retail distributor in Toronto called Mike and Mike's that went to extreme lengths to promote our products and support our farm. They really taught us how to sell vegetables to retail customers. Mike and Mike's had an insatiable appetite for cucumbers. "Give us all the cucumbers you have, every week," one of the Mikes told us. "We'll buy whatever you can grow."

Every small farm has a complex mix of variables that constantly interact with and influence each other. Soil type, climate, size, labour model, location, level of mechanization, markets, crop mix—even the personality type of the people who work there. When one variable changes, there are repercussions for all the others. We had gradually changed our crop mix, which eventually led to drastic changes in our labour model, which had allowed us to serve different markets and grow our sales, but it also forced us to rethink the seasonality of our farm.

The small farm orthodoxy is very big on season extension. Eliot Coleman, who wrote the book that provided the blueprint for our farm, has also written several books on season extension. He and others advocate the use of unheated hoop houses and cold frames to plant crops earlier and harvest them later, which allows a farmer to extend the growing season and thus make more money. We had put in a root cellar to store our beets and potatoes and thus extend the season of our root vegetables, but after a year with *los muchachos*, we had to reassess. Our crew wanted two things: to work as many hours as possible and to be away from their families for the shortest possible time. So rather than extend the season, we had to compact the season. We had to cram as much production and sales as possible into a short growing season so our guys could get the hours they wanted, then get back home. This had the added benefit of making our life more pleasant. Washing beets in January is one of the shittiest jobs I have ever undertaken. In Season Seven, the guys showed up at the end of April, when Gillian and I had already done the first few plantings, and left at the end of October. When the guys went, we shut everything down and were done.

IN SEASON SEVEN WE MADE ANOTHER CHANGE, one that had nothing to do with maximizing production or increasing profitability. We decided to stop raising pigs. Our pigs had never been a money-maker for us. We raised two or three every year to feed our family, to use for events on the farm and to give to friends and relations. We sometimes traded pork for lamb with Gord and Teza at the Art Farm. But the pig-butchering weekend in the fall had become a bit of an institution—we would kill, scald and gut the pigs on Friday, then put out an open invitation to our chef clients to help with the butchering on Sunday. Creemore Springs always gave us a keg of beer, Gillian cooked a big lunch, and a good time was had by all (except, of course, the pigs).

The annual pig kill strengthened our relationship with Carl and Ryan at Marben Restaurant, in particular. They came every year to help on the kill day, and Ryan always supervised the butchering. The most accomplished chefs, some of them very skilled butchers themselves, always deferred to Ryan when it came to cutting; he was a master. In 2012 Carl won *Top Chef Canada* and used his prize money to create a new restaurant, in partnership with his soon-to-be fiancée, Julia, and Ryan. When Richmond Station opened in Toronto's financial district, it was an instant hit. Carl ran the kitchen, Ryan oversaw the in-house butchery and charcuterie, and Julia ran the front of house. They often brought their whole team up to the farm to learn about how we produced our food, and simply to have fun. The people they hired were pretty much all of a type: very young, ridiculously talented and genuinely nice.

The last year we raised pigs, Carl and Ryan came up with a couple of young sous-chefs to help with the kill. The first pig was a large female. I led it into the barnyard with a bowl of food. Ryan had asked to kill the pig that day; he was ready with the .22. The pig buried its face in the bowl and Ryan lowered the rifle into position. Just as he pulled the trigger, the pig jerked its head up. There was a loud *crack* from the rifle, but instead of dropping, the pig froze, then started squealing. It was a single, piercing, high-pitched scream, the pig rooted in place, blood pouring out of its mouth. Ryan froze. I grabbed the rifle from him, chambered another round, and dropped the pig with a bullet to the brain. Then Carl jumped in and stuck it. The time between Ryan's shot and mine must have been less than two seconds, but it felt like an eternity. We were all shaken. Ryan still talks about that day; even as a professional butcher, that brief moment of limbo between the botched shot and the *coup de grâce* made him think about his role in the killing and eating of animals in ways he hadn't before.

It made Gillian and me think about our role, too. The way we felt the first time we killed our own pigs, the feeling that what we were doing was cruel and barbaric, only intensified as the years went on. Every year we saw more and more evidence of our pigs' intelligence and good nature. Our dog, Zip, has a long and shaggy coat and often gets burdocks buried deep in his fur. He would sometimes present himself outside the pigpen, and one pig would stick its head through the fence and gently pull the burrs out of the dog's coat with its mouth. We also grew increasingly aware of the catastrophic role that animal agriculture plays in climate change. It didn't matter if we looked at it from the perspective of ethics or the environment or our own health; it became ever more difficult to figure out why we were eating so much meat. Our only justification was that it tasted really good, which is a pretty weak justification for killing an animal.

Foster and Ella had grown up with the idea that the meat they ate came from animals they knew, and they never had a problem with that. From a very early age they would name all the pigs each year, then sit around the dinner table in the winter and speculate about which pig had produced the chops they were eating. I wasn't sure if that was a good thing or a bad thing. Gillian and I finally made the decision to stop raising pigs, and we decided to stop eating beef altogether. We still raise some meat chickens and we eat a lot of eggs, but overall we eat way less meat than we once did. The chefs miss the annual pig kill and the kids miss the chops, but eating less meat is quite simply an unqualified good.

SEASON SEVEN, OUR SECOND SEASON with *los muchachos*, ended very differently than all the others. We didn't have any pigs to kill. We sold out of beets and potatoes as fast as we pulled them out of the ground, so there was nothing to put in the root cellar. Gillian and I were exhausted, but we weren't as physically broken as we had been in the past. In late October I drove the guys down to the airport, we hugged and said goodbye, and then I drove home to a farm that was already closed for the season, completely shut down.

When Gillian and I sat down for our annual end-of-year meeting, we already knew that, from a financial standpoint at least, it had been a breakout year. With an experienced crew and the right equipment and infrastructure, we were able to increase our production while maintaining extremely high quality. The skill and competence of our employees freed Gillian and me from having to be in the garden all the time, and gave both of us (but mostly Gillian) time to sell all that extra food we were producing. Our increased production had allowed us to go into retail in a big way and to serve new and bigger customers. If in Season Six we had reached a tipping point, then in Season Seven we went over the edge. We had achieved a scale that just made everything click. In August we broke our single-day sales record: over nine thousand dollars. We sold more in that one day than we had sold in our entire first season.

When we crunched the numbers, we realized that we hadn't just turned a profit, we had paid off everything—the farm, the schoolhouse, the solar panels and all our equipment, including the new hoop houses and irrigation system. We were 100 percent debt-free. We had set out seven years earlier to create a small-scale sustainable organic farm that fully supported our family. And we had finally done it.

CHAPTER 11

THE STOP

FROM THE VERY BEGINNING, our farm faced a conundrum: we wanted our food to be accessible to as many people as possible, but we couldn't afford to sell it for anything less than top dollar. We hadn't started the farm with the intention of becoming purveyors of salad to the one percent, but lowering our prices wasn't an option. We knew how much money, time and effort it took to grow our produce, and in the beginning we were still losing money hand over fist. If we were serious about making our food more accessible, we would have to be creative.

It was Gillian's parents who provided the first clue to solving our accessibility-versus-income conundrum. Remember way back in Season Two, when we had a short-lived plan to start a CSA? I said earlier that we sold only one share, but that wasn't quite true. Gillian's parents also sent us a cheque for a share, but since they were way down in Vermont they suggested we donate it to a local food bank. We liked that idea so much that we suggested other CSA members buy a second share to donate, a plan we called "share a share," but since our CSA died before it began, the idea didn't go anywhere.

Dick and Kathi wouldn't take their money back, so Gillian and I decided to donate five or six hundred dollars' worth of food to the closest food bank, in Stayner. That's when we started learning a lot of uncomfortable truths about food banks. The organization in Stayner was typical of those in many small towns—run by a handful of elderly volunteers on a shoestring budget out of a borrowed space. They were open only a few hours a week and didn't have capacity to store or distribute fresh produce. We approached the food bank in Collingwood, run by the Salvation Army, but got the same story. The volunteers in both places were incredibly dedicated and caring individuals, but they didn't have any resources to work with. The food they were giving out when Gillian went to see them represented the dregs of our industrial food system—highly processed, unhealthy, unappealing and cheap. Back in 2008, that summed up the offerings in the vast majority of food banks all across North America.

We never would have guessed that it would be so difficult to give away fresh food, but we persevered. It was about this time that Gillian started attending a lot of local food events in Toronto. Among the chefs and farmers and local food activists at the events, she kept running into a guy named Nick Saul. Nick was executive director of The Stop Community Food Centre, but he had worked for years before that as a community organizer and political agitator. He also happened to be married to Andrea Curtis, a writer and activist who was an old friend of mine. Nick, tall and slim, looked a bit like George McFly, only cooler.

The Stop had been a typical food bank when Nick took the helm years earlier, doling out supermarket cast-offs and boxes of macaroni and cheese from a cramped, dingy location. Nick quickly became disillusioned with the food bank model. He and others at The Stop came to believe that handing out cheap, unhealthy food to people who couldn't afford any better served to reinforce and institutionalize poverty rather than address it. People couldn't afford to buy food because they didn't have enough money. The food bank was merely keeping them alive, facilitating an economic and social welfare system that didn't pay people enough to feed themselves. The only way to solve the problem of hunger, they reasoned, was to confront the root causes of poverty. The real genius of The Stop was to realize that food could be a catalyst for this broader change.

The Stop was housed on the ground floor of a public housing building on Davenport Road, in Toronto's west end, in the middle of one of the poorest, most underserviced neighbourhoods in Toronto. It had installed a large commercial kitchen and set up a dining room with big windows and lots of light. Rather than lining up and meekly accepting their hampers, food bank clients would pick out the fresh produce and healthy food they wanted from a room set up like a store. The Stop hired Joshna Maharaj, a trained chef who had worked in some of the city's best restaurants, to run the kitchen and prepare free meals for community members. Everything The Stop did revolved around the idea of dignity, from the physical surroundings to the attitude of the volunteers to the quality of the food that was offered.

This philosophy is best summed up by a story Nick often tells. A big part of The Stop's work is advocacy—training clients and community members to lobby for policy change to address poverty. When Nick and his staff met with politicians to talk policy, they often brought community members along. Nick once brought a woman named Nicole to a meeting with a provincial cabinet minister. An immigrant from St. Lucia, Nicole had started out as a Stop client when she was a twenty-five-year-old single mother of three; The Stop had given her the support and community she needed to train as a social worker, and she had recently graduated from community college. It was December, and the minister started the meeting by telling everyone about one of her Christmas traditions. "Every year I take my children to the supermarket," she told the gathering, "and we all buy food to donate to our local food bank." Nicole looked the minister in the eye. "There is no dignity in giving my children what your children want to eat," she said quietly.

By the time we first called Nick, The Stop was running several community gardens, offering cooking classes and pre- and postnatal nutrition programs, and providing hot, delicious lunches to hundreds of people a day. It was also helping clients access all kinds of other services in the community—affordable housing, mental health services, daycare, you name it. Nick was unreservedly enthusiastic about getting our food into The Stop's programs. "Send us whatever you've got," he told Gillian. She called Paul and Grace at 100km Foods, and they immediately agreed to help. "We'll deliver to The Stop for free," they said, without hesitation.

Early in Season Three we packed up our first shipment to The Stop, back when we still met the 100km truck at Barrie Hill. I can't remember exactly what we sent—at that time of year it must have been salad and radishes and maybe some spinach. What I do remember is that we took great pains to ensure that the stuff we were sending to The Stop was every bit as good as the stuff we were sending to the restaurants that day. We took the same obsessive care in harvesting, washing and packing everything, so the bins going to The Stop were indistinguishable from those going to Jamie Kennedy or Canoe or anywhere else. I always feel a lot of pride when we send off a load of particularly beautiful produce to a client, but I felt a little something extra that day.

In two weeks we had exhausted all of Dick and Kathi's donation. Joshna, who ran the kitchen, called us up to tell us how our product had been received. "I just served two hundred people the most beautiful fucking salad I have ever seen!" she raved. "They loved it. I want more!" Gillian called Nick and told him that we would be willing to donate product if they could use it. Nick was adamant in his refusal. "We can't expect small farmers to shoulder the financial burden of fixing our broken food system," he said. "You people have to get paid, or you won't stay in business." So we scrounged and scraped and found ways to keep our vegetables flowing. The Stop used some of its own money to buy from us, and we solicited donations from our friends, family and market customers. When we told a retired couple about our initiative in the market one day, they wrote us a cheque for a thousand dollars on the spot. 100km always delivered for free. We often donated surplus product, but Nick wouldn't accept a delivery unless at least some of it had been paid for. At the end of Season Three we sent an email out to our community, asking people to help us "Grow for The Stop," and the name sort of stuck.

THE GROW FOR THE STOP MODEL that we developed with Nick had two explicit and complementary goals: to give people in low-income communities access to the best organic produce available, and to support small-scale organic farmers. Nick's insistence that we get paid for our produce turned our accessibility-*versus*-income problem into an accessibility-*and*-income solution. If we were going to build an alternative food system, we needed to make good food accessible and to make small farms profitable; Grow for The Stop was intended to do both. It started as a tiny step, but at least it was a step.

In the beginning, the income side of the Grow for The Stop equation benefited only one farm: ours. The amount of money we were able to raise was very small, but so were our total sales, so

The Stop quickly became an important client. By Season Four, the second year of our partnership, The Stop accounted for more than 10 percent of our total sales, more than any single restaurant. In those lean early years, selling to The Stop helped keep us afloat.

By Season Four, the scale of the need had also become starkly apparent. The Stop's clients by then numbered in the tens of thousands, an indication of both the success of their model and the enormity of the problem. When we asked Nick if they could use more of our food, he looked at us like we'd been smoking something. "We're basically a bottomless pit," he said. They were serving thousands of people a week in the community kitchen and food bank, and were still turning people away. So we decided to throw a party.

I can't remember exactly how we came up with the idea, but early that summer we found ourselves preparing for a big fund-raiser on the farm. Our friends and market customers had been generously supporting the Grow for The Stop program, but we needed more and we didn't think we could go back to that well again. Our friend Sara, who ran the *Creemore Echo*, had started promoting concerts on the side, booking acts into the little community halls in Avening and Singhampton. She had been working with a guy named Fred Eaglesmith, a crusty old singer-songwriter from Port Dover, on the north shore of Lake Erie, who often played two hundred shows a year. Fred had a small but fanatical following of "Fred Heads" who sought out his shows as he wandered around small-town North America. Sara booked him for our event.

As soon as we started planning the concert, volunteers came out of the woodwork. Matt Flett, who had sold his tomatoes with us in the market, volunteered to supervise the food. Matt was a culinary instructor at the community college in Barrie and an accomplished chef. Creemore Springs Brewery offered to donate all the beer. Sara volunteered to organize the permits. Everyone in Dunedin offered to help us run the show. Jim and Dan volunteered to help with set-up; they built a "disco bale"—a straw bale wrapped in Christmas lights—that they hung high in the rafters of the barn. When you're throwing a party in our community, it's never hard to find help.

Sara made up some posters, we told all our friends, and we sold tickets at the *Echo* and at our stand in the market. Gillian cold-called a few companies that seemed to be jumping on the local food bandwagon to see if they might sponsor the event. Bernardin, the company that makes home canning equipment, agreed right away. Before they even met us, they sent a cheque for five thousand dollars and enough jars to serve all the beer. We bought half an organic pig from Gerald te Velde, and when the big day arrived, Matt and the interns cooked up a huge feast of pulled pork in a big Cuban barbecue box that Matt had bought for the occasion. The crowd was a bizarre mix of the different communities we had come to be part of: weekenders in button-down shirts, CRAFT farmers in overalls and plaid, tattooed interns, Stop staff and community members, and a whole bunch of middle-aged Fred Heads, who must have done this type of thing before, because every one of them brought a lawn chair. Our friends from Dunedin arrived on an armada of ATVs. There were about 250 people in all. Everyone milled around the big bonfire in the barnyard, drinking jars of beer and eating pulled pork sandwiches and basking in the beautiful June evening. Having so many people enjoying themselves on our farm made me absolutely euphoric, a feeling I wasn't prepared for, given my usual aversion to people invading my space.

Nick made a speech and then Fred took to the makeshift stage in the barn. He played all his hits, including "I Shot Your Dog," and several numbers about trains. I got drunk and Gillian and I abandoned all pretense of being organized or in charge. Everything just sort of happened, and everyone was happy. It was a beautiful, magical night.

WE HAD SO MUCH FUN THAT NIGHT that we decided to do it again in the fall. This time we started with a play, a one-woman show about the founding of a Mennonite community in northern Ontario, that held the crowd of 150 people spellbound on their straw-bale seats in the barn. The band that night was called the Sunparlour Players, a banjo-driven punk/country Mennonite trio that sounded a little bit like the Violent Femmes. They sold their homemade jams and preserves at the merchandise table. The two events together raised almost twenty thousand dollars, all of which went to The Stop in the form of fresh organic vegetables from our farm.

Our sales to The Stop were important in the early years, but holding big fundraising events on the farm was an extremely labour-intensive way to generate business. Most of the organizational burden fell on Gillian, who was a master of coordinating volunteers, booking Port-O-Lets and cajoling sponsors. From a purely financial point of view, selling to The Stop didn't really make sense: it was just too much work to raise the money. But in the early years farming itself didn't make sense from a purely financial point of view, so we kept on doing it.

Our business did benefit from our association with The Stop in myriad other ways, however. For starters, it introduced us to a whole community of activist chefs. The Stop had done an excellent job of recruiting chefs to their cause, using culinary star power to raise funds and awareness. Nick invited us to Stop fundraisers, where we met chefs such as Bertrand Alépée, Matty Matheson and Chris Brown, who would later leave the restaurant business to work for The Stop full-time. Working with The Stop was like a badge that identified the true believers—chefs who recognized the social, political and environmental significance of food. We were often the only farmers at the events we attended, and the chefs took note. "Okay, you get it," they seemed to say to us. "You're one of us." Many of them became our clients.

Working with The Stop also gave us a huge advantage when we were recruiting interns. By Seasons Four and Five, there were a lot of CRAFT farms and a limited pool of prospective interns. Without a CSA or horses, we had to find other ways to attract good candidates. For many young people interested in farming, social justice and accessibility were the primary motivators. Prospective interns didn't just want to learn how to farm, they wanted to change the food system. We highlighted our partnership with The Stop in all our advertisements and sent our interns down to the city for a week during the summer to work in The Stop's community gardens and kitchen. For many of our interns, that was the highlight of the whole experience. Without the draw of The Stop, we might not have filled all our positions.

At the Season Five fundraiser, Matt Flett, Gillian and the interns cooked again, and we had a band called Elliott Brood in the barn; they were a lot younger, a lot better known and a lot louder than Fred. I thought they were going to blow the barn down. They played one of my all-time favourite songs for their encore—a cover of Neil Young's "Powderfinger"—while I drank rye from a bottle side-stage with the crew from one of the other CRAFT farms. It was perhaps the greatest musical moment of my life.

By Season Six, word had started to spread in the Canadian music community that we put on a pretty good event. That year we landed Stars, an indie rock band that sells out venues many, many times larger than our barn. They agreed to do the show for free. When a power cable blew mid-show, the crowd went silent. Amy Millan, Torquil Campbell and the rest of the band walked off the stage into the middle of the barn and sang one of their songs a cappella. Those who were there still talk about that moment.

That year we also invited some of our chef clients to cook at the event. We ended up with some of the best chefs in the country on our front lawn, a dozen in total, all volunteering their time and cooking at stations over open fires. Carl, Ryan, Julia and the crew from Richmond Station ran the backstage in our house, cooking for the band and the sponsors in our little kitchen.

Bringing so many chefs to our farm was another transformative moment for our business and another way that our relationship with The Stop benefited us. We had had a few chefs come up to tour the farm independently of our Grow for The Stop events, but now we had dozens. Chefs are a lot like farmers. They work very long hours, often for little money. Their work requires a combination of manual dexterity, strength, endurance and intelligence. They have the dual imperative of creating and selling beautiful food. And, like farmers, they are driven primarily by a love of their craft. Our chefs all loved cooking on our farm, in sight of where their ingredients were growing, for a crowd that deeply appreciated good food. Chefs also like to party. Our fundraisers transformed many of them from clients into friends and fellow food crusaders. Our business relationships grew into something deeper and more enduring. We all recognized that we were working toward the same higher purpose.

The chefs who cooked at our Grow for The Stop events started coming up to the farm at other times of the year. John Horne, the executive chef at Canoe, brought up his entire staff—cooks, dishwashers, front of house, the works. We took them on a tour and showed them how to plant with the EarthWay; they helped weed the lettuce and then we had a big barbecue lunch. Carl and Ryan brought up the staff from Richmond Station every winter for a day on the farm. We cooked outside, ate and drank, and played pick-up hockey on the rink I made on our front lawn. Carl was in his mid-twenties back then, and his staff were almost all very young. One year Foster corralled a bunch of cooks into a prolonged video game session. I was about to admonish him for monopolizing their time, but then realized that they were all much closer in age to Foster than they were to me, and were probably all very happy to be on the couch with him. The food was always a highlight when chefs came to visit. Ella and Foster both loved it when the young cooks drafted them into service as sous-chefs for the epic meals they prepared over our front patio campfire or in our kitchen.

OUR PARTNERSHIP WITH THE STOP introduced us to new chefs, helped us attract interns and built intense loyalty with our chef clients, but the biggest impact it had on our business was helping us to create a successful retail brand. The typical small farm marketing channels—farmers' markets, CSAs and restaurants—can form the basis of a sustainable business, but the vast majority of food is still sold in supermarkets. Small farms face big challenges when trying to sell to supermarkets.

The first issue is the supermarkets themselves. The problems with the big chains that dominate the sector are well understood; suffice it to say that they are massive corporations that are focused almost exclusively on the bottom line. They demand huge scale, year-round supply and rock-bottom prices. A friend who runs a large organic operation tells the story of showing up at a meeting with a buyer from a large chain and being presented with a box of tissues. "You're going to need these when I tell you how much we're going to pay you for your produce," the buyer said. We have never had the scale to work with the big supermarket chains, and we don't like selling to assholes, so that option was out.

Even with small grocers and independent retailers, price is a challenge. Grocers typically find that they have to mark up produce 30 percent or more to cover their fixed costs and the inevitable spoilage and still leave room for a narrow profit. When we added up our wholesale price and markups for our distributor and the retailer, it was simply more than people would be willing to pay. The labelling and packaging required for retail products is also extremely expensive when bought in small quantities. One simple elastic tag for a bunch of carrots cost us thirty cents, and we had to buy a minimum of five thousand. The more research we did, the more insurmountable the obstacles appeared. But Gillian eventually figured out a way to crack the retail nut.

The Stop had long been supported by a couple of progressive independent supermarkets in the city: the Big Carrot in the east end and Fiesta Farms in the west, in our old neighbourhood. The Big Carrot was all organic, and Fiesta had transformed itself from a traditional Italian grocery store into a champion of fresh, local food. Nick Saul introduced Gillian to Joe Virgona, the elderly, gregarious, not very politically correct owner of Fiesta Farms, at a Stop fundraiser in the city. Joe wanted to carry our stuff but couldn't figure out how he could pay the price we were asking and still make money. Gillian came up with the idea of creating a Grow for The Stop retail brand.

The concept was pretty simple. We would sell all our retail produce under a Grow for The Stop label. The stores would donate 10 percent of the retail price back to The Stop, which would use the money to buy local organic produce for its programs. We came up with promotional materials with lines like "When you buy these vegetables, you feed your family, and your community."

Consumers would feel good about themselves because they would be helping increase access to organic food, as well as supporting local farmers. The retailers would benefit from being associated with The Stop, a well-known and progressive local organization. The retailer would be willing to make a little less and the consumer willing to pay a little more, so we could sell our produce at a price that allowed us to make money. While much of the money that we raised at our Grow for the Stop events went to buying food from our farm, we made a deal with The Stop that all the money generated through our retail sales would be used to buy food directly from farms other than ours, so that other small farms could start building a relationship with The Stop. Everyone got something out of the deal, and the numbers worked for everyone.

We started out selling bunched beets and carrots to the Big Carrot and Fiesta Farms, and from the beginning, almost all our cucumbers went into retail stores. The big retail breakthrough came when we started selling bagged salad. We had resisted bagging salad for years, mostly because we hated the thought of using that much plastic. We asked the Carrot and Fiesta over and over to please offer our salad in bulk, so people could load up their own bags with tongs, but they refused, saying that their customers simply wouldn't go for it. We knew that with interns we would never have the time to bag salad, so we shelved the idea.

As our retail sales increased, Gillian spent more and more time on the phone with retail produce managers, especially Carmen at Fiesta Farms and Chris at the Big Carrot. They were both lovely people, and they taught us a great deal about the ins and outs of the produce business. Carmen finally called us up one day. "You realize," he said, "that even if we sold your salad in bulk, our customers would still put it in a bag." He had a point. By then we had our Mexican crew and we were actively looking for ways to increase their hours, so we had labels made up, sourced a micro-perforated salad bag and started bagging. The response was overwhelming. We went from nothing in Season Six to over eighty thousand dollars in retail salad sales in Season Seven. We picked up Mike and Mike's, our retail distributor, and our cucumber sales also went through the roof. Expanding into retail was an important factor in breaking through to profitability, and it wouldn't have happened without The Stop.

IT WAS JUST AFTER SIX ON A SATURDAY MORNING in July when I woke up to the sound of a trailer banging down our laneway. I got up, threw on some clothes and walked outside. It was sunny and dry and already hot; it was going to be a scorcher. Matt Flett was backing his SUV into the far corner of our big front yard. I walked over and helped him unload the Caja China—the Cuban barbecue box that he had bought for our very first event. It was extremely heavy. Matt lifted a corner of the lid to reveal a half-dozen huge pork shoulders, all coated in a thick, dark rub. "I'm going Korean this year," he said with a wink, pointing to a massive glass jar of homemade kimchi in the trailer. Matt no longer did all the cooking for our events, but he came every year to make some variation on pulled pork. His station was always one of the most popular, despite being surrounded by chefs from some of the best restaurants in the world. I helped Matt get the fire going on top of the box. His pork would slow-roast for twelve hours before the first guests arrived.

It was 2015, our Season Nine, and our annual Grow for The Stop fundraiser had grown far beyond what it was in the early years. Our old friend Patrick, who had first taken us to Runaway Creek Farm in Quebec all those years ago, runs a music management company, and he helped us book bands that really had no business playing in our little barn. After Stars we had the Sam

Roberts Band, who were some of the nicest guys you could ever hope to meet. They put on a show that still stands as the greatest display of straight-up rock 'n' roll I have ever witnessed. The year after that we had Gord Downie and the Sadies, who had released an album together. Gord had such a good time that he vowed to come back the next year with his band. He was true to his word; much to my amazement, the Tragically Hip would be playing in the barn that night.

Now, American readers might not grasp the magnitude of the talent that ended up on our makeshift stage in the barn. The Sam Roberts Band is a stalwart of the Canadian music scene, held in the highest esteem by critics, fans and fellow musicians alike. The Hip is even bigger—they are without a doubt the most beloved band in the country, and have been for the past twenty years. It's not a stretch to say that the Hip playing in our barn was akin to Springsteen showing up at a backyard barbecue in the States.

The culinary side of the event had also grown over the years. From a barnyard buffet cooked up by Matt and our interns, the food had morphed into a dozen stations on the front lawn, run by chefs from some of the best restaurants in the country. We had Amanda Ray from Biff's, Jason Bangerter from Langdon Hall, Giacomo Pasquini from Vertical, John Sinopoli from Table 17, Michael Robertson from Oliver & Bonacini—the list goes on. Scott MacNeil, who had taken over The Stop's kitchen from Joshna, took his rightful place among the culinary heavyweights every year and demonstrated the kind of delicious, dignity-building food that he creates every day for The Stop's community members. A number of great restaurants had sprung up in our area in the years since we started farming, and they were represented, too: places like Azzurra in Collingwood, Bruce Wine Bar in Thornbury, and Creemore Kitchen down in the village. Richmond Station always ran the backstage for the band and sponsors, which was a big reason why we kept getting such great bands. That night, for the Hip show, we had many of the usual suspects cooking on the front lawn, as well as Sam Gelman and Paula Navarrete from Momofuku, Sylvain Assié and Jennifer Dewasha from Café Boulud, and Jamie Kennedy. The crowd that night could be divided between star-struck foodies who were there for the chefs, rabid Hip fans who were there for the band, and a whole lot of people who were there for both.

All day, chefs loaded in and started cooking, volunteers put up tents and lugged bales of ice, and the guys from the brewery unloaded dozens of kegs into our walk-in cooler. Foster turned fourteen that summer; he was in charge of preparing the bonfire. Ella was twelve; she helped set up the merchandise table. All our friends from the community had been helping out with the event for so long that they needed little direction. Sara from the *Echo* still helped with the organizing, Tara from Dunedin supervised all the volunteers, and Gillian oversaw the whole operation, a pillar of organization in a chaotic sea of happy, busy helpers. My job was to hang around in case anything heavy needed to be lifted. A tent city rose down by the irrigation pond. Most of the chefs and a lot of the volunteers planned to camp out overnight.

Shortly after noon, the roadies arrived, a big truck and a tour bus towing a trailer full of gear. They had driven overnight from a show in Ottawa. Amps and speakers and soundboards and cases of guitars were loaded into the barn. When the sound check started, I got the feeling I always get on the day of the show: butterflies in my stomach, happy anticipation, excitement. It's the best day of the year.

A few hours later, the band arrived. They parked their tour bus in the field behind the barn, out near the hoop houses. They had finished a very long North American tour the night before and were all glad to be off the bus and out in the sunshine. They wandered around the farm, chatting with the volunteers and checking the set-up. Then they all ate a beautiful meal prepared by Carl and Ryan at our dining-room table in the house, family style.

The guests started arriving shortly after five. Tickets that year had sold out in a matter of seconds, and we had maxed out our capacity. We sold over 750 tickets, but with chefs and volunteers we would have close to a thousand people on site. I had mown a parking area out in the field beyond our farthest growing area, so people would park and walk in through the gardens on their way to the barn. By the time the people started coming, Gillian and the volunteers had everything under control and there was nothing for me to do but fidget and pace around the lawn, so I went out and helped direct traffic in the parking lot, like I always did. The mayor of our township and his buddy Rolf, who sells us our firewood, arrived with their huge campers and set up a trailer park in the field. Two different groups from Collingwood came in chartered school buses. The few Dunedinites who weren't working as volunteers arrived on their ATVs, as usual. Juan Carlos, Chuen, Luis and the rest of the guys walked over from the schoolhouse; we always gave them the day off for the event so they could relax and enjoy themselves. I handed them a fistful of beer tickets. Soon there was a river of people snaking through our garden, walking past the places where the food they were about to eat had been harvested a few hours earlier.

By six o'clock the front yard was full of people drinking craft-distilled gin cocktails, organic Southbrook wine from Niagara and jars of Creemore Springs beer. There was also a lot of serious eating going on. The evening sun filtered through the huge maple trees and the clouds of woodsmoke from the cooking fires. Just about everyone whom I loved in the world was there in one place, eating great food, drinking great booze and thoroughly enjoying themselves. I was deeply and profoundly happy.

There was a pause after dinner so the chefs could break down their stations before the show started. Then the crowd squeezed into the barn and spilled out into the barnyard. Foster lit the bonfire and the band took the stage, light streaming through the cracks in the barn wall behind them as the sun set in the west. Ella led a bunch of girls to the very front, at the edge of the stage, as she always did, and Foster sat on top of the stone foundation wall at the side with a bunch of boys, their feet dangling in a row.

I can't describe what it was like to hear all those iconic Hip songs played in my own barn, surrounded by all my best friends and family members. They played a long set, with several encores. The sound was amazing and they were incredibly tight—a legendary band at the height of their powers. At least five different people approached me during the show, put their hands on my shoulders and shouted above the din. "Dude!" they yelled in drunken disbelief, "The Hip are playing *in your barn!*" I couldn't believe it either. The feeling that had come over me when I first walked into that barn twelve years ago, the feeling that we had found our place, suddenly made sense.

Toward the end of the show I found Gillian in the crowd and grabbed onto her and held her as tight as I could. This was all her doing, a result of her energy and talent and generosity. The vibe, the joy, the palpable feeling of collective happiness that night was something that she had conjured up, with the help of an army of beautiful, talented, generous people. How had I been lucky enough to end up with someone who could do something like that? "You did this," I whispered in her ear. "Thank you."

CHAPTER 12

SMALL IS BEAUTIFUL

UNTIL THIS POINT, I've tried to stick pretty closely to the chronology of our farm story, bringing you along on a journey from an apartment in downtown Toronto to a full-blown, profitable organic farm on top of a hill near the little village of Creemore. The story has followed a reasonably tidy arc—a couple of idealistic urbanites and their kids leave the city to start a farm, they struggle and make mistakes and learn, and eventually they succeed. It would be nice to think that the end point of the story is a happy embrace in the midst of a huge party with a legendary band playing the soundtrack. But that's not really how the story goes. We didn't succeed at farming and then live happily ever after. We succeeded at farming and then realized that financial success wasn't enough. We struggled for almost a decade to reach our goals and then realized that the goals we'd been chasing weren't the ones that really mattered. This epiphany came about a year *before* the Hip show in our barn. So now we need to disrupt the chronology and rewind the story a full year. There weren't any crowds around. There wasn't a band playing. It was just me and Chuen, out in the beet patch.

Chuen is the middle brother between Luis and Juan Carlos. He is kind and generous and just about always cheerful. Chuen loves music; he has a massive digital collection representing all of the many Latin American genres—banda, cumbia, salsa, and a lot of really shitty Mexican pop. He's also a pretty good barber; Chuen had a side business selling playlists and cutting hair, back when he worked on farms with large crews of Mexican workers. The Perez-Victoriano brothers have very different personalities. If I wanted one of the brothers to run the farm while Gillian and I were away, I would pick Juan Carlos. If I wanted to go out on the town for a night of partying, I would take Luis. But if I had to be stranded on a desert island with one of the brothers for an extended period of time, I would choose Chuen. He's just a genuinely nice guy.

It was August 2014, Season Eight on the farm, our third season with our Mexican crew. The year before had been our breakout year. We had sold almost three hundred thousand dollars' worth of vegetables and had paid off all our debt; we were finally profitable and financially sustainable. Gillian had stopped taking on consulting contracts a few years earlier, and neither of us was doing any work off-farm. We had achieved what had seemed impossible eight years earlier: we were making a living on the farm.

But that day out in the beets, talking to Chuen, I felt much as I had on any other August day since we had started the farm. I was anxious, stressed out and exhausted. We had taken on more workers and opened more land that season; we now had a crew of six and twenty acres of garden. We had distributors picking up five and sometimes six days a week. The pace had only become more frenetic, and we were locked in a constant struggle to keep up with demand. Business-minded friends looked at our operation and saw nothing but opportunity. If we could make good money with six guys and twenty acres, why not make *great* money with a dozen guys and forty acres? We had the land, the experienced staff, the distribution network and the know-how—why not cash in? But just thinking about it made me tired.

I don't remember exactly what Chuen and I were talking about that day, but it had something to do with Morris, from Barrie Hill. Chuen had worked for Morris for a number of years, and they were close. Chuen was telling me a story about something Morris had done on the farm, and I made some sort of favourable comparison between me and Morris. "Oh, no, no, no!" Chuen interjected with a laugh. "You and Morris are very different. Morris is always happy, not like you!"

He said it cheerfully, without a hint of malice. *Morris is always happy, not like you.* Chuen clearly felt that this was obvious and self-evident. I felt like I had been stabbed in the chest.

Chuen's words cut me so deeply because, I realized, they were true. On the surface, I had every reason to be happy. Our farm business was thriving. Famous chefs and high-end retail stores were lining up to buy our produce. We had started selling to progressive hospitals and universities. We couldn't keep up with demand. Gillian and I had been asked to speak at conferences and symposia, sharing the stage with global culinary superstars. We had brunch with René Redzepi and Magnus Nilsson. We hung out with Ben Shewry. Gillian took Mark Bittman out for dinner when he was visiting Toronto. When Daniel Boulud and David Chang launched restaurants in Toronto, they sought us out and began buying from us the day they opened. Best of all, we were actually making money. We had created a profitable, sustainable organic farm, from scratch.

I reflected on all these accomplishments as I walked in from the field that day and felt a profound sense of disappointment. So what? I asked myself. I had focused so narrowly on building the farm that I had neglected everything else in my life. I had drifted apart from my friends and family. I had completely abandoned my intellectual life. I was often impatient with my kids, and I resented shuttling them to their activities because it interfered with my work. I had come to see Gillian, with whom I spent almost every waking hour, more as an antagonistic business partner than as my beautiful, caring wife. I spent eight or nine months of every year working like a dog, and the rest sitting isolated on a frozen farm. From a financial perspective, the past eight years had been a modest success. But Chuen was right; I wasn't happy. That's really the only perspective that matters.

THAT FALL, IN EARLY OCTOBER, my mother died. Her lymphoma, which seemed to have disappeared for six or seven years after she was first treated, had come roaring back. Gillian and I had picked out the most beautiful food from the farm to bring to her when she got sick again, but by early summer she had pretty much stopped eating. My parents had left the city and moved to Collingwood a few years earlier. Now Gillian drove the half-hour to their house several times a week to lead my mom in the gentle yoga that seemed to be the one thing that made her feel better. When she became too weak for yoga, Gillian lay with her in bed and did breathing exercises with her. My dad, my brother and sister and I were all with her when she left us. When I came home to the farm that morning, the guys came immediately to embrace me. They were all crying.

I had taken my mom to one of her last chemotherapy treatments at Sunnybrook Hospital in Toronto the previous June. I was shocked by how busy the place was. There were hundreds of patients hurrying in and out of the reception area. It seemed like everyone was going through a familiar routine, with no sense of urgency or dread. The atmosphere of the hospital made cancer seem normal, which I found terrifying. My mother received her treatment in a large outpatient "chemo suite" where half a dozen people sat in easy chairs, hooked up to IVs. The nurses called out each patient's name and date of birth before turning on the drugs, to double-check that they had the right person. The woman receiving treatment next to my mom was born in 1969, the same year I was.

My mom died at the busiest, most frantic time of the year on the farm, when I'm usually the most stressed and worn out. But that year, my mind was obviously elsewhere. I was forty-five years old, and it suddenly occurred to me that I had almost certainly lived more than half my life already. My mom died at seventy-one; who's to say I won't go even sooner than that? Life's too short to spend in a state of perpetual stress about the next harvest. It's also too short to waste it chasing the wrong dream.

LATER THAT MONTH, A FEW DAYS AFTER Juan Carlos, Chuen, Luis and the others flew home to Mexico, Gillian and I sat down for our end-of-season planning meeting. It was cold and wet outside, like it always was, and we had a fire going in the woodstove, like we always did. Over the years, I had come to dread those meetings. The fundamental problem was that Gillian and I had never resolved the growth-versus-survival conundrum. Gillian correctly asserted that we had a financial imperative to grow and sell more. I correctly asserted that we were already exhausted and overworked. That year, as we gathered our papers and settled into the couch by the fire, I braced myself for the inevitable fight. We looked at each other in silence for a while, neither one wanting to start.

"What if . . ." Gillian began tentatively. "What if we didn't grow the farm next year? What if we planted the same amount and just made a few tweaks to make things more efficient?" I sat there, slightly stunned. Our numbers were great. We had sold even more than the year before and we had spent way less. Without any debt to pay down or big infrastructure to buy, we had managed to pay ourselves a half-decent salary. It was only about half the amount that we had been making a decade earlier, when we had real jobs in Toronto, but it was enough to satisfy pretty much all our needs. It wasn't a huge amount of money, but it was real.

I had expected Gillian to be all fired up by our financial success. I thought she would want to really cash in, to use our momentum to get our salary up to what it would have been if we had stayed in the city, rather than half of what it had been ten years before. It never occurred to me that she would suggest we stop growing. We had so far measured our success almost exclusively by how much we had grown each season, by how much more we could produce. Growing was what you did when you were in business. That's the way the world worked. Not growing struck me as a radical idea.

"What if," Gillian continued, "we focused instead on the things that really make us happy?"

And with that, everything changed.

For eight years, our end-of-season meeting had been about how to make more money. Now we spent the day dreaming of ways to make a difference, to make change, to make more happiness. We decided to build a commercial kitchen and event space on the farm to bring chefs, foodies and farmers together to plot a new future for agriculture. We decided to redouble our efforts to make good food accessible to all. We decided to make the creation of wealth and opportunity for *los muchachos*, our employees, a central goal of our business. And we decided to enjoy what we had already achieved, rather than always looking to the future and worrying about what's next. It was a beautiful, life-changing day. A world of possibilities opened up to us. I couldn't stop smiling.

GILLIAN AND I HAVE MADE more than our share of stupid mistakes over the years, and, like everyone, we have lots of flaws. But we do have one thing going for us: when we make a plan, even a crazy one, we follow through. Over the past two years, all the crazy ideas we came up with that day at the end of Season Eight have been put in motion. First, we expanded the profit-sharing scheme we had been running with our employees for several seasons. Our guys are the biggest factor in the financial success of the farm, so they should benefit directly from that success. Almost all of our workers have used the extra money to start businesses back home in Mexico. Luis always drove a taxi in his home village during the off-season; last year he bought his own taxi and hired a driver. Chuen bought a bus to carry people back and forth to Mexico City. All of their kids are in school. Juan Carlos's son wants to be an engineer.

We put even more effort into our Grow for The Stop events and retail brand. The show with the Tragically Hip raised almost $75,000 in one night. This year Sloan played in the barn, and our fundraising total exceeded $100,000. The Stop still buys more of our produce than any single restaurant, and 100km Foods still delivers for free. Last year we started supplying a new community food centre located in Regent Park, a maze of public housing complexes in downtown Toronto that's one of the poorest neighbourhoods in the city. We also bought organic vegetables, fruit, cider and honey from a half-dozen local farms to fill four hundred Christmas hampers for the Collingwood and Stayner food banks.

Our friend Jamie has built a thriving construction business in the years since we sold with him in the farmers' market. In 2015 he started building our event space, which we call the New Farm Kitchen. We decided from the start that the purpose of the new space wouldn't be to make money, which had the curious effect of making people want to give us money. The brewery in Creemore gave us a bunch of cash toward the construction costs, as did Bernardin, the home canning company that had sponsored our Grow for The Stop events from the beginning. Almost all the gear for the kitchen was donated. When you remove the profit motive and do something because it's fun and cool and important, everyone wants to help.

The New Farm Kitchen opened in the spring of Season Ten, with a fully kitted-out commercial kitchen, space for twenty-five to dine, a patio and outdoor fireplace, and a couple of bunk rooms for chefs to crash in. The building is on the cutting edge of green construction; it's hyperinsulated and gets almost all of its winter heat from the sun shining through the big south-facing windows. Now we host pop-ups with visiting chefs, fundraisers for progressive farming organizations, canning workshops and field trips for culinary students at the local high schools. Nick Saul left The Stop to found Community Food Centres Canada, an organization dedicated to spreading The Stop's model to other communities. Nick now brings committed young chefs from all over the country to our farm for media and advocacy training, helping them become spokespeople for the good food revolution.

The more we paid our employees, the more money we raised for other organizations, the more we built places for hanging out and entertaining our friends and community members—in short, the more we focused on things other than making money—the more financially stable and profitable our business became. Our annual sales are now close to half a million dollars, and we still plant only about twelve acres a year (the rest is in cover crops). Last winter I got a call from a researcher at Statistics Canada who was conducting an annual survey of fruit and vegetable growers. He ran me though a bunch of questions about the different crops we planted.

"How many acres do you have in potatoes?" he asked.

"About four," I told him.

"What about lettuce?"

"A little less than five."

"Cucumbers?"

"Less than one. We only grow them in our hoop houses."

"Wow, you guys are really small," he said. He was used to talking to large conventional growers. "What would you say were your gross sales per acre last year?"

I thought about it for a minute. "About forty thousand dollars an acre," I told him.

I heard a lot of keystrokes in the background. "Are you sure?" he asked. "The form on my computer won't accept a number that large." He tried and tried, but he couldn't figure out a way to enter our data. It seems that the survey's designers had never contemplated any farm selling forty thousand dollars' worth of food from a single acre. We had done it with nine people, some hand tools and one small tractor, and no pesticides.

PEOPLE OFTEN COMMENT ON HOW LUCKY we've been; we started our farm at almost exactly the same time that interest in local and organic food began to explode. Then we started selling to chefs when "farm to table" was just becoming a thing. Then we started selling into retail as the market for organic moved from the fringes to the mainstream. We rode the wave of interest in good food to financial success on our farm, but here's the important thing: there were lots of others riding that wave with us. There were thousands of other small farms like ours that "got lucky" at the same time we did.

In the ten years since we started our farm, we have witnessed broad and profound changes in the way our society looks at food. Just think of all the words and phrases you probably never heard a decade ago—food miles, food deserts, foodies, farm to table, urban agriculture—and you start to realize how much has changed. As more and more people began to recognize the problems with the industrial model of agriculture and the Western diet, more and more people changed their relationship with the food they ate. They changed what food they bought and where they bought it. They changed their expectations of the restaurants they dined in, which changed the position, status and influence of chefs. They became willing to pay for attributes that might not be readily visible on a supermarket shelf: how the food was produced, where, and by whom. For many people, food has come to occupy a central place in their self-identity, the defining attribute of their lifestyle or politics. "I am a foodie." "I'm vegan." "I am a locavore." All of these changes taken together add up to what we, and many others, call the good food movement.

The good food movement is still young and messy and ill-defined. There are a lot of disparate people on board—hedonists and environmentalists, jet-setting chefs and hippie homesteaders—and sometimes the whole thing can seem a little precious. The movement is still too much about personal choice and not enough about collective political action to bring about real policy change. But one thing is clear and undeniable: the good food movement has made farms like ours possible.

Gillian and I worked incredibly hard for a decade to make our farm a success. We did some things that I think were pretty innovative. We recognized that we had to put as much effort into selling our food as we did into growing it. We kept meticulous records and made business decisions based on hard data. We focused relentlessly on quality, producing vegetables that were worthy of the best restaurants on the continent. And we made real efforts to make our food accessible to everyone. But the crucial factor in our success, the thing that put us over the top, wasn't anything that Gillian or I did; it was the good food movement itself. Without the wave of people willing to pay more for good food, no amount of hard work or innovation would have put us where we are today. Have you seen those bumper stickers that say "Farmers Feed Cities"? "Those should be paired with another sticker that says "Cities Employ Farmers." It's people who care about good food—people like you—who literally make it possible for people like me and Gillian to farm.

The changes that have made our farm possible have made lots and lots of other small farms possible, too. I talk to farmers almost every day who are riding the same wave we are, seeking out new markets and opportunities, carving a niche in the new food economy. These farmers tend to be young, educated, often female, and fiercely political. They are well aware of the fact that they're bucking an industrial system of agriculture that would be more than pleased to see them fail, but they're motivated and persistent, intent on forging a new model. Gillian and I tend to rail against the small farm orthodoxy, but some of our fellow farmers are doing extremely well by it, running sophisticated and profitable operations based on farmers' markets and CSAs. Others are doing what we have done, selling wholesale to restaurants, grocery stores, home delivery services and more. There are farmers making money with pick-your-own, agro-tourism, on-farm food processing—you name it.

There are as many marketing strategies as there are small farms, but what all successful small farms share is the ability to innovate. We're all riding the wave of the good food revolution, and none of us knows exactly where it's going; we have to be nimble and creative and ready to switch gears at a moment's notice. Small farmers are the ultimate entrepreneurs, running vertically integrated businesses that must do everything well, from primary production to sales, marketing and customer service. We take seeds, and with hard work, creativity and intelligence, we make something beautiful and valuable and essential for human life.

The challenges small farmers face are real and daunting, and they differ on either side of the U.S./Canada border. The organic movement in Gillian's home state of Vermont is light years ahead of ours, and organic growers in the States generally enjoy far more government support than we do. But universal healthcare and subsidized tuition give Canadian farmers (and other small businesspeople) a huge advantage. I graduated from college with only three thousand dollars in debt, and we have never even had to think about health insurance. High student debt or worries about health coverage probably would have kept us anchored at our desk jobs in Toronto all those years ago.

Despite the challenges, what we need now are more small farms, and more farmers. What Gerald Poechman told me all those years ago is more true now than ever: the rest of agriculture is "hurtling in the opposite direction." If we want to make serious change in the way our food is produced, if we want to do more than tinker at the edges, we need more farmers. A lot more.

There's no way we could have established our farm without the advice, wisdom and encouragement we received from other small farmers. A fellow CRAFT farmer once put it to me this way: "Our competition isn't other small sustainable farms," she told me. "Our competition is conventional agriculture." She was right. Everyone in the community of small-scale organic farmers recognizes that the revolution requires many, many more people to join our ranks, so just about everyone is generous with their knowledge. Gillian and I try to repay the help we received in the early years by being completely open and transparent about our practices and strategies. We often host tours for other farmers where we disclose everything: the varieties we grow, the equipment we use, the customers we sell to, even our financial information.

We may have stopped growing our own business, but the market for the kind of food we produce is growing in leaps and bounds. We need more farmers to meet the demand.

Our society's current obsession with all things food and farming isn't an unalloyed good. It seems like a lot of people spend more time watching people cook on TV than they do actually cooking. Big Food is busy finding ways to co-opt the movement, subtly adopting language that evokes small-farm agriculture. I found a McDonald's flyer in our mailbox a few weeks ago that proudly proclaimed they served only "farm-raised" chicken. As opposed to what? Our focus on local food also overlooks an inconvenient truth: industrial agriculture is everywhere. No matter where you live, there are local farmers who are poisoning their land with pesticides, growing nothing but genetically modified crops and raising their animals in conditions of unspeakable cruelty. Farmers like those shouldn't be part of our movement, no matter how local they happen to be.

But on the whole, the changes we've seen over the past ten years have been overwhelmingly positive. The good food movement has created conditions where farming sustainably on a small scale, for the local market, is now a financially viable option, not for just a few people but for lots of people. The message to all you foodies out there should be loud and clear: Keep on doing what you're doing. It's working.

WHEN GILLIAN AND I DECIDED to stop growing the farm, we knew that we would have less material wealth as a consequence. What wasn't fully apparent at the time was how much richer our life would become. We were deliberately trading financial capital for social capital, giving up money for time, friendship and love. That trade has turned out to be the best we ever made. We've been the recipients of a bonanza of social capital, and the farm has become a very happy place indeed. Sometimes I think I'm the luckiest guy in the world.

Our farm is big enough that we can make a decent living but small enough that it's manageable and fun. It takes me three and a half minutes to walk from my back door to the farthest part of the garden, yet that garden provides everything our family requires. Our entire operation is visible from our back porch, but it generates enough money to satisfy all our needs. It is immensely satisfying to live surrounded by such a compact, self-contained family support system, to know that our little patch of garden provides a livelihood not just for our family but for the families of all our workers as well. I can't for the life of me figure out why we would ever want to get any bigger. I think we've found the sweet spot.

I get to live and work in the midst of a stunningly beautiful landscape. I often step out my back door in the morning and catch my breath when I see the sunrise. On any given day I might see a bald eagle or a whitetail deer or a coyote running by as I work. I can hike and cross-country ski on the Niagara Escarpment for hours and hours and never come upon another human being.

Gillian and I get to work with people of great talent, passion and commitment—chefs, food activists, fellow farmers and the beautiful people of Dunedin and Creemore—all of whom value friendship and sharing a great meal above money and material wealth. They're always willing to chip in and help. None of them will ever give a shit what kind of car we drive. We are all bound together in a collaborative community, and everyone is willing to share freely. It's our community that gives us licence to devote our energy to things that are more fulfilling and important than simply accumulating cash.

Our children get to grow up with fresh air and clean water and great food. They have been foraging in the garden since the time they could walk, eating cherry tomatoes, raspberries and even heads of broccoli right off the plant. They now have the confidence and independence of farm kids. Foster has been driving since he was twelve. Ella cooks elaborate meals from scratch without ever looking at a recipe. Gillian and I get to start each day by having breakfast with our kids, and we're both there every afternoon when they get off the school bus. Sometimes I worry that our kids are too isolated on the farm, cut off from other kids and experiences, so in the winter we take them out of school and we travel. We've been to New Zealand and Europe, and a few years ago we spent six weeks road-tripping through eastern Africa, showing Foster and Ella all our old haunts. We might not have as much money as our former colleagues in the city, but none of them travel like we do, because we have something more valuable: time.

And now that I am no longer constantly striving for more, no longer always thinking about what's next, I can enjoy what I do each day and appreciate the value of my work. The kind of work I do is still physically hard—I lift and pull and carry and dig. My work makes me hungry, which makes good food taste better. My work makes me tired, so I sleep well at night. I can see the literal fruits of my labour every day, which gives me a great deal of satisfaction. But my work is also intellectually challenging and stimulating. It takes a lot of thought to keep a crew of seven employees working at peak efficiency. Gillian and I have to make decisions based on thousands of ever-changing variables, and manage chaotic and unpredictable natural processes in order to create a predictable and uniform output of extremely high-quality food. We deal with hundreds of clients and thousands of people who eventually eat our produce. When people describe farm work as drudgery, I laugh. For me, working in a cubicle was drudgery. Farm work is constantly changing, mentally stimulating and, above all, fun. I wouldn't want to do anything else.

The work we do is tangible and tactile. We can see and touch the products of our labour. Gillian and I are the masters of our domain, and we exercise unfettered control over our business. We make the decisions and we reap the rewards, or suffer the consequences. We are competent. We have agency. The work we do is good and important.

Farming, in short, isn't just a good way to make a living. Farming for us is the foundation of a meaningful and happy life. It's the ultimate expression of independence and self-sufficiency, but it's also a deeply political act, one that has the power to build community and inspire others to action.

What better way to live your life?

VIVA LA REVOLUCIÓN!

TEN YEARS AGO, when we first contemplated the idea of starting a farm, I was looking for a grand cause, a project big enough to make my life's work. I never could have guessed at the time what a grand project food would turn out to be. For more than a decade, Gillian and I have wrestled with the most practical and intimate realities of building a sustainable organic farm from scratch. But at the same time, we have always tried to think about what we're doing on our little farm in the context of the wider food system, and in the context of the struggle for positive social, environmental and political change. Our conclusion, after all these years, is simple: food is everything. Food is at the centre of all of the biggest, most important, most pressing challenges we face as a species. Food is the unifying theory of all that is wrong in the world, but it is also the key to solving our biggest problems.

That may sound like a sweeping statement, but I think it's accurate. Take, for example, human health. Diet-related illnesses are no longer a problem just in rich countries; they now kill more people globally than anything else. Our industrial food system is making us fat and sick, and as the cheap, processed, sugar-laden staples of the North American diet get exported around the world, the problem is only getting worse. According to the British medical journal *The Lancet*, obesity now kills three times as many people as malnutrition.

Or take inequality. The way we eat is both a symbol and a symptom of the staggering inequality of North American society. Our food system mirrors our economic system: the rich get organic and the poor get diabetes. As the gap between rich and poor widens, the gap between what the rich and poor eat also gets bigger, to the point where the gap between how long the rich and poor live is growing again for the first time in more than a century.

What about social connection and happiness? For all of human history, across all cultures, eating has been an opportunity to gather, talk, connect and bond. But our atomized, homogenized food culture threatens to erase that vital tradition. As we break down into households of one, scatter far from our families and forget how to cook for ourselves, we lose the simple pleasure and vital social function of the shared meal. Breakfast cereal sales have been declining for years because breakfast cereal is not convenient enough, meaning it can't be eaten in a car. Just think of it, all of us sitting alone in traffic, eating breakfast bars. It's so sad.

Or think about a big, sweeping problem like environmental degradation. Every human activity has an impact on the biosphere, but growing food causes environmental mayhem on a scale far beyond anything else we do—habitat destruction, deforestation, chemical contamination, mass extinction, eutrophication of lakes and rivers. We are on the verge of destroying the global fishery, and many of the world's great forests have been cleared to make way for farms. It's no exaggeration to say that all our biggest environmental problems are really food-system problems.

Or take the biggest problem of them all: climate change. The way we eat has a bigger impact on our climate than anything else we do. Some analysts have calculated that food and farming generate more than half of all greenhouse gas emissions, and that animal agriculture alone produces more greenhouse gases than the entire transportation sector. When it comes to climate, what you put on your plate probably has a bigger impact than what you park in your garage.

Our industrial food system is destroying our environment, hijacking our climate, making us fat and sick and unhappy. It's a huge, overwhelming, complex, multi-faceted problem, but the good news is that there is a huge, complex, multi-faceted movement underway to fix things. Right now, all over the world, farmers and activists and chefs and ordinary people are harnessing the power of food to confront the big, intractable problems that we face. Idealistic young people are growing food in poor urban neighbourhoods. Organizations like The Stop are fighting inequality and loneliness one meal at a time. Farmers are producing beautiful, nutritious food in a way that enhances rather than destroys their environment. Eaters are making a new system possible simply by choosing to buy good food. All of these people are actively and deliberately building an alternative food system, each and every day.

Gillian and I have always believed that the foundation of the good food movement must be farms that are both sustainable and profitable. The great news is, that nut has been cracked. Our farm, and many others like it, prove that an alternative model is viable, that you can farm in a way that enriches the soil, cleans the atmosphere and strengthens community, and not just get by but make a good living. We no longer need to invent a new model; the model exists. It's working. Replicating that model, scaling it up and expanding it, will take time. There are lots of other problems with the food system that also need to be addressed. But small sustainable, entrepreneurial, committed farmers are laying down the bedrock of a new system. If we build on that foundation, anything is possible. The good food movement is powerful. The revolution is underway.

A decade ago, Gillian and I set out on what we thought would be a straightforward quest: to create a profitable, sustainable organic farm. That quest soon came to consume us. It broke down our bodies. It threatened to destroy our marriage. It was, by far, the hardest thing either of us has ever done. It took a very long time and it cost us a great deal, but eventually we succeeded, and what we ended up with is so much more than we ever could have hoped for. Our farm provides us with everything we need: great food, clean water and enough money to do pretty much whatever we want. We have time to spend with our children and a safe, stimulating environment for them to grow up in. We live surrounded by a nurturing and caring community. Our work is satisfying, meaningful and important. Our farm has become a place that inspires chefs, activists and artists alike. Our home is full of passionate, creative and committed people.

Our farm is not a fluke or an exception. There are many others like it, and the forces that allowed our farm to succeed are only getting stronger. The revolution needs cooks and organizers and distributors and businesspeople and eaters and a whole lot more farmers. What role will you play?

ACKNOWLEDGEMENTS

THIS BOOK, LIKE OUR FARM, would not exist without the generosity and assistance of many, many people. For our farm, the list of those who helped us along the way is far too long to include here, but without the support of a great many chefs, farmers, retailers, distributors and good-food activists, we never would have made it. A number of those who had the greatest influence on us are mentioned in the book. For those who are not, our gratitude is no less sincere. We owe a special debt to our friends and neighbours in Dunedin and Creemore; it was the desire to be and remain part of this remarkable community that was our primary motivation for persevering on the farm.

I have the good fortune to count three professional writers as friends, all of whom offered advice and encouragement at the outset of the book-writing process: Dan Clements in Dunedin, Andrea Curtis in Toronto and Duff McDonald in New York. Duff in particular was unfailingly generous in helping me craft the book proposal. My agents, Cindy Uh and Meg Thompson, not only treated this project with care and commitment, they helped me turn our ten-year ordeal into a coherent story. Amanda Lewis and

Anne Collins, my editors at Random House Canada, made the book better with each draft. I am especially grateful to Jamison Stoltz at Abrams Press for his enthusiastic embrace of the book, and for helping it to reach a much wider audience.

My greatest debt is owed to my friends and family, who have supported my every endeavour. I had long and helpful conversations about the book with many friends, but the longest and most helpful were with Mark Mullen and Graham Moysey, both of whom read early drafts and provided valuable suggestions. Steve McDonald drew a map and provided artwork for the cover and would accept only vegetables as payment. My family greeted the book idea the same way they greet all my quixotic schemes, with love and enthusiasm and a lot of good-natured ridicule. Thanks to my sister Lindsey and my brother Mike and their beautiful families; to my in-laws, Dick and Kathi, who have been our most valuable source of farming advice and unflagging supporters; to my father, who always said I should write a book; and my mother, who really believed that I could accomplish anything. I wish she could have read our story.

It seems strange to acknowledge my immediate family, because this story is as much theirs as it is mine, but I'll give it a shot. Ella and Foster, you are the motivation for everything we do on the farm and in our community. This book is for you. Gillian, you are my greatest ally, my keenest critic, my truest partner in our farm, this book and everything else we do together. Thank you for joining me on this wonderful ride.

BRENT PRESTON worked as a human rights investigator, aid worker, election observer and journalist on four continents before finding his true calling as a farmer. In 2003 he and his wife, Gillian Flies, abandoned successful careers in Toronto, packed up their two young children and moved to a rundown farm outside Creemore, Ontario. Since then, they have built The New Farm into a thriving business and a leading light in the good food movement, providing organic vegetables to some of the best restaurants in Canada, and raising hundreds of thousands of dollars to make good food more accessible in low income communities. Preston speaks often on food and farming issues and writes for the *Huffington Post*.

www.thenewfarm.ca